GLOBAL ENERGY TRANSFORMATION

GLOBAL ENERGY TRANSFORMATION

FOUR NECESSARY STEPS TO MAKE CLEAN ENERGY THE NEXT SUCCESS STORY

Mats Larsson
Management Consultant

First published 2009 by
PALGRAVE MACMILLAN

Palgrave Macmillan in the UK is an imprint of Macmillan Publishers Limited,
registered in England, company number 785998, of Houndmills, Basingstoke,
Hampshire RG21 6XS.

Palgrave Macmillan in the US is a division of St Martin's Press LLC,
175 Fifth Avenue, New York, NY 10010.

Palgrave Macmillan is the global academic imprint of the above companies
and has companies and representatives throughout the world.

Palgrave® and Macmillan® are registered trademarks in the United States,
the United Kingdom, Europe and other countries.

ISBN-13: 978–0–230–22919–8

This book is printed on paper suitable for recycling and made from fully
managed and sustained forest sources. Logging, pulping and manufacturing
processes are expected to conform to the environmental regulations of the
country of origin.

A catalogue record for this book is available from the British Library.

A catalog record for this book is available from the Library of Congress.

10   9   8   7   6   5   4   3   2   1
18   17   16   15   14   13   12   11   10   09

Printed and bound in Great Britain by
CPI Antony Rowe, Chippenham and Eastbourne

Dedication

To my wife Bodil and my daughters Cajsa and Maja with thanks for all their support through the writing of this and previous books.

Global Energy Transformation Institute, www.getinstitute.com

Contents

Acknowledgments

I would like to thank my friend Stuart Pledger, who helped with the development of the idea for the book in the early stages and who also contributed to the climate change discussion and to the part on his method for collaborative problem solving, the "Collaboration Café." Thanks also to Julian Darley and Richard Heinberg of The Post Carbon Institute in California. The e-mail questions you asked during the writing of the book generated ideas such as "the global embrace" and added to my thoughts on organizational learning in society. Peter Wennström, whom I met by chance at Schiphol airport, enthusiastically contributed ideas for the title of the book. Mikael Edholm, at the Swedish Trade Council, contributed his extensive knowledge about prediction markets.

Thanks to Thomas Björkman, Johanna Moberg, Glenn Widerström and Urban Bergström at the Swedish Energy Agency for support in the development of ideas on energy improvement of supply chains. Thanks also to Håkan Knutsson and Per Simonsson at Sustainable Business Hub. Our discussions gave me the ideas that generated the part about communication of ideas that drive behavior change toward sustainability.

I also thank Hans Enocson, Nordic General Manager of GE, for generously giving his time and contributing information about the GE Ecomagination project. Similarly, Inge Horkeby, Director at Volvo AB, in a series of meetings over the past 2 years, explained to me the opportunities of energy transformation in the transportation sector.

Harry Frank, former Research Director at ABB and Professor at the Royal Swedish Academy of the Sciences, provided early insight in the need for leadership in the process of energy systems transformation. Many thanks also to Lennart Swanström, Senior Scientist at ABB, and Pontus Cerin, Assistant Professor at the Swedish Environmental Institute, for contributing their knowledge about life cycle analysis.

I wish to thank Eva Ståhl, Fredrik Adolfsson and Thomas Bergmark, managers at IKEA, for the generous contributions of information about this company and its projects in the areas of energy and the environment.

Thanks to Bengt Gebert, Jenny Bramell, Kjell Persson, Curt Schröder, Ingela Bogren and Åsa Dahl for daily encouragement and support.

I am also grateful to the team at Palgrave Macmillan led by Stephen Rutt for their support and ideas.

CHAPTER 1

Executive Summary

1.1 BACKGROUND

The transformation of existing energy systems to sustainable and renewable alternatives represents a tremendous management challenge. We need to view it as an opportunity to renew energy systems and renew important parts of the economy and social life. In addition to this we may see it as an opportunity to create new business and, if we manage the process well, we may also see it as an opportunity to create new economic growth. The magnitude and complexity of this challenge has not been widely recognized. So far, primarily the technical aspects of different alternatives have been debated. We know a lot about "what" has to change. So far, however, very little has been written about the "how" of the change process.

We have access to a number of renewable technologies that are currently available on a small scale, but few of those are available on a large scale. In most cases available technologies need to be improved or new technologies need to be developed, production resources for technologies, products and fuels need to be expanded and we also need to expand the knowledge base in terms of the number of people who have competence in both the areas of technology and change management during the transformation. The speed of all activities related to this change needs to be increased.

The magnitude of the challenge is indicated by a few examples:

- The global production of oil amounts to 85 million barrels per day.
- If all US cars and trucks were to be fueled by grain-based ethanol, all land in the country would have to be used in order to grow grain and there would be no land left to live on and no grain left to eat.[1]
- If all US cars and trucks were to be fueled by electricity, 500 new nuclear plants would be needed. In the whole of the world today, there are some 400 plants in use. If the cars were to be driven on hydrogen fuel cells, substantially more plants would be needed, since it takes more electricity to make hydrogen through electrolysis, than to run the cars on electrical batteries directly.[2]
- If all European cars and trucks would be run on cellulose-based biofuels, 1000 large plants for the production of biofuels would be needed and each would need deliveries of wood by 450 trucks per

day. European forests would not be able to supply more than a fraction of the necessary raw material on a sustainable basis.[3]

- In order to expand the production of wind power so that 12 percent of global needs are supplied by wind power by 2020, we need to build 630,000 new wind turbines from 2002 until 2020, each with a capacity of 2 megawatts. On average this means that we need to add 70 gigawatts per year of new capacity during this period. In 2002 the global production capacity for wind turbines was 7 gigawatts per year. This capacity had increased to some 20 gigawatts per year in 2007, but the global capacity is still far behind schedule if we aspire to achieve the lofty goal of 12 percent wind power by 2020.[4]
- Other renewable energy technologies, such as photovoltaic power and wave energy, are still at an early stage of their development compared to wind power. These technologies are currently not competitive without subsidies and the total amount of electricity generated from these sources is minuscule.

In addition to this a substantial share of nuclear plants in Europe and the United States are aging and they are approaching a point where they either need to be replaced by plants with new, higher capacities or need large-scale refurbishment, which sometimes could be coupled with capacity increases.

In order to achieve the goal of sustaining and improving global prosperity, while facing the need of large-scale energy transformation, an energy transformation program needs to take advantage of the most important experiences of previous large-scale programs, and cutting-edge knowledge about change management, regardless of where in the world this may be located. In this book we draw on experiences from a number of large-scale American change programs, but we also use industrial examples from both European and American companies, and there are also experiences from Asia and other parts of the world indicating that energy transformation is already a global effort. In the future, the global component may even need to be strengthened through even stronger ties for technology and knowledge transfer between transformation programs in different parts of the world.

1.1.1 The Drivers behind Change

Energy systems transformation is not only a technical issue but also a high-level management, financial and political issue. There are two new drivers that force us to take decisive action. The most readily quantifiable of these, from a business and economic perspective, is the "peak oil" issue. Recent research indicates that we are approaching the global peak in oil production and that it will arrive sooner than previous estimates have indicated. Global economic growth is still strongly correlated with increases in energy consumption, even if a number of Western economies have broken this relationship. We are, however, dependent on growth in the global economy for European economic growth to continue. If we want economic growth to continue we need to rapidly develop new energy systems and transform existing utilities and industrial and transportation systems to the use of renewable energy sources.

We also have the pressing issue of global warming, which is altering living conditions and economic conditions at an alarming rate. The economic signals coming from the consequences of climate change in the direction of the economic system are in some ways weaker and less clear than those from peak oil. Therefore it may be difficult to increase the levels of economic and market-based activity to levels that are necessary, based on the signals from climate change alone.

1.2 THE FOUR STEPS THAT ARE NECESSARY

This book identifies four steps, that are necessary in order to transform global energy systems. These are the following:

1. **Analysis** of the overall situation and the contributions that different solutions could make to solving the problem will be necessary. Interrelationships between solutions and the implications of these need also be analyzed. The analysis must focus on change management, technology, financing and other important aspects of the change. This, on global, regional and sector scales, as we will see, is a massive, but necessary undertaking.
2. **Strategy development** globally and in each country is necessary for the transformation. We need a high-level map of the general direction of change, which also puts in place the role of market incentives and the level of government action in different areas. A strategy is also necessary for each of the sectors that are identified in the book as the key to transformation: transportation, utilities, industrial processes, buildings, agriculture and behavior and work life change.
3. A number of **plans** must be developed based on the strategies. Plans need to be available on different levels and they all need to correspond. There needs to be an overall plan for the transformation in each country and sector and more detailed plans for particular projects and programs within sectors.
4. **Managed change** shall be the key to successful transformation. The market forces could not be relied upon entirely to produce precise results at high speed. Governments need to manage change in their countries and in joint efforts on a global scale. Descriptions

of large-scale US projects in the past serve as examples that support this argument.

Up until now the debate has centered on the idea that markets and financial incentives based on market economics need to determine which investments and activities should be undertaken. Some experts, however, such as Congressman Jay Inslee and Bracken Hendricks in their book (which has a foreword by former President Bill Clinton) *Apollo's Fire*, argue that a planned program similar to the American Apollo program that put a man on the Moon in 1969 needs to be put in place.

This book is based on a similar argument, arguing that markets are relatively slow and imprecise as tools to achieve rapid and focused action. We need a program of planned activities in combination with market-based action, in order to arrive at a successful outcome in the short period of time that we have at our disposal. Governments and companies need to become active in the high-level strategic analysis and planning on the overall level and it cannot be excluded that substantial financing and other types of intervention and incentives are needed from governments all over the world.

Box 1.1 A Very Simple Example That Illustrates the Situation

Energy systems are extremely complex. It makes sense to reduce this complexity by describing a relatively general situation, which could serve as an example of a typical situation for energy use and savings opportunities all over the world. It is relevant both for small businesses and for large companies. To work as an illustration of the latter the reader only needs to think of many users instead of one in the example below.

A person, let's call him Mike, runs a distribution business. Every morning he receives goods from four different producers, which he uses his small truck to distribute during the day. On a typical day he loads his truck 80 percent full in the morning, drives one round of customers and arrives back home at noon. After lunch he once more loads his truck 80 percent full and delivers to the other half of his customers. He leads a reasonably good life, having the opportunity to take his family on vacation in summer and eat good food at home.

Suddenly he finds that the price of diesel has gone up by 30 percent over the course of 3 months, which increases his cost of operations by 5 percent. This makes him a little worried, because his profits decrease. It has very little impact on his lifestyle however, since he manages to pass on some of the cost increase to his customers.

During the next 6 months the fuel prices continue to increase and Mike finds that he better improve the efficiency of his business. Having a small business he cannot afford to assign a consultant the job or buy a piece of computer software for optimizing routes, so he sits down himself with his map every evening trying to figure out the best route to drive the next day, given the orders at hand. This takes him one hour every evening, but he finds that the time is well spent, because it reduces his fuel cost by 10 percent and he manages to load the truck up to an average of 90 percent in the morning and do the extra 10 percent deliveries on the same amount of fuel that he has always used for the morning round.

Now he hears on the radio that the price increases are because of the reduced supplies of crude from oil-producing countries all over the world. All users of petroleum products are urged to find ways to reduce consumption. He thinks that he has already made significant savings on his own, but he knows that many of his neighbors and friends have not, so he hopes that they will now start to calculate in the same way that he has done. On the radio they also say that the price of electricity is related to the price of petroleum, so people with factories that use electricity for heating and for running machines are urged to try to find ways to save. He realizes that many factories could easily save certain amounts of energy simply through better planning.

Now, he starts to hear among his friends that they have started savings activities similar to his own and he sees that the price increases have started to taper off. That is a relief!

The result of this is only temporary, however. Mike hears on the radio that supplies continue to decrease and that there may be shortages at the pump. Now he starts every day by filling up, so as to have fuel for the next week, in case there may be shortfalls in delivery. He spends 30 minutes queuing, since other people have thought of the same. Now he realises that, in case the situation gets worse he will not have enough fuel to serve all his customers. Even with the well-planned routes that he is now doing every day, he needs to go 150 miles per day on his routes. In case he does not get enough fuel, he will need to cut down on his deliveries and serve fewer customers.

He decides to buy a hybrid truck, which reduces fuel consumption by 30 percent. This truck, just as his present truck, could also run on biodiesel, which increases his options when shortages occur. The dealer tells him that, although he has been planning well in advance, the production of hybrids is still so low that there is a 2-year waiting list for a new one. The truck producer is about to expand capacity however, and the increased capacity may come on-line in 18 months. This may reduce his waiting time to less than 2 years. The dealer warns him that it will become difficult to trade in the present vehicle, because of the very low demand for trucks with high fuel consumption. All customers are asking for trucks with low fuel consumption. This is a disappointment and a financial loss for our friend, since the truck is only 8 years old and, techni-cally, has a number of years' use ahead of it.

Mike tells his friends and neighbors about this. Some have tried to buy new electricity-efficient machines to replace their existing ones and the waiting lists are between 2 and 4 years in all cases. He realizes that only a few percent have tried to order replacements. Many could not afford to replace their old trucks and machines and others have not started to work on the problem. The informa-tion that he has received says that only 12 percent of demand for trucks could be satisfied by hybrids and trucks that could run on renewable fuels. Production needs to be expanded rapidly in order to replace existing truck fleets. The same situation exists in most markets for machinery and vehicles. At the same time, production of biodiesel and electricity from renewable sources needs to expand in order to keep up with the rapidly increasing demand. And this is just to maintain the present levels of economic activity. In order to make economic growth possible, even larger amounts of renewable fuels need to be made available at short notice.

The two things that impact the progress of this development are broadly the level of oil reserves and the level of mitigation activities that have been in progress throughout the development. The people with access to better overall information need to help our friend and his neighbors in the example to forecast his future needs and to plan production in advance in order for business and society to be able to satisfy them. In order for enough hybrid trucks, cars and trucks using different biofuels and other energy-efficient and sustainable vehicles and machines to be available when people start to demand them in larger quantities, efforts to increase production of them need to be made well ahead of the expected increases in demand.

1.2.1 Planned Programs in the Past

Three examples of large-scale planned government programs were dis-cussed, all of which were based on different levels of planning and government action. When the US industry in 1942 was transformed into a war economy, all resources in the country became focused on one goal. This was the goal of making the United States into "the arsenal of democracy" and to produce everything necessary in order to win the war on the part of the Allied forces. The Apollo program was run in parallel with normal economic activities and the Marshall Plan was a financial and management effort to help in the rebuilding of Europe after the Second World War. None of these planned economic programs has made the US economy into a planned economy. The characteristics of planned economies in socialist states that include state ownership of production resources and lack of incentives for individual initiatives have never been present in any of these programs, and they will not be part of any planned effort in the future either.

1.2.2 Government-Financed Technology Development Is Necessary for Economic Growth

Contrary to the belief of many politicians, economists and business people, government investments have been necessary for economic growth in the past. Market-based activities have only been complementary to government-financed long-term technology development. This is the conclusion of Professor Vernon W Ruttan, who studied the development of six different general purpose technologies and found that investments by the government in military technology development have been necessary in order to develop the American production system, airplane and computer technologies, the Internet, space technologies and nuclear power. In the case of nuclear power, Ruttan argues that this technology, most probably, would never have developed in the absence of government-funded programs. In the cases of other technologies, development would have slowed down substantially.

Governments have the ability to invest in high-risk and long-term technology projects that are required to develop new general purpose technologies. Companies generally do not take the risks of decades of investments in new technologies. When we consider photovoltaics, wave energy, fuel cell or electric vehicles and a number of other technologies, we are looking at long-term development projects.

Governments need to analyze the challenges, opportunities and prospects of technology development in a number of sectors of society and the economy and make overall plans for action in the sectors of

- transportation,
- utilities,
- the built environment,
- industrial processes,
- agriculture, and
- behavior and work life change.

Governments in each country also need to determine the level of planning and government financing needed. This analysis needs to be made in cooperation with companies that are involved in different technology areas and technology and management experts in different fields.

Plans need to be made for the overall, aggregate level of society and for each of the sectors above. Based on these plans, projects need to be initiated that solve different technical and management problems in the same way that there were technical and management projects within the Second World War transformation and the Apollo program.

Plans need to be based on an understanding of the limited resources that need to be used in the transformation program. The most important of these are the following:

- time,
- money,
- raw materials,
- land, and
- competence (technical, financial, economic, administrative and management).

In Part 2 of this book a number of tools for the management of the change process are provided and, in Part 3 technical and transformation opportunities in each of the six areas from transportation to behavior and work life change are discussed in some depth, mainly from a change management perspective. Case examples are provided illustrating the opportunities and the challenges offered by the different alternatives.

The situation in each of the sectors is summarized very briefly here. In all areas, financially justifiable investments need to be prioritized. When necessary investments are not financially justifiable or very risky for single companies, possible financing solutions need to be analyzed and debated. Sometimes public financing may be required or innovative financing solutions may have to be developed.

1.2.3　Transportation

In transportation we are heavily dependent on oil. There is currently no renewable alternative that is available short term on a large scale. Biofuels, fuel cells and various electricity-based alternatives all need further technology development and expansion of production resources. Raw materials and logistics systems for these are important issues that need to be planned for and dealt with. Transportation systems, vehicle and fuel production and distribution are all large-scale systems that are interrelated in a number of ways, and strategic political decisions need to be made in order to reduce risk in technology development, implementation and expansion.

1.2.4 Utilities

There exist a few solutions that could be expanded, but the lead times for expansion are long. Existing alternatives are nuclear (10 years' construction of new plants), cogeneration and wind power. Promising new technologies such as solar photovoltaics and wave power need further development to become cost-competitive. Tidal power, wave power, wind power and nuclear power have different types of environmental issues related to them for permits to build plants or for safety and storage of residues. The democratic issues need to be handled on the level of society, and alternatives that are selected for expansion need efficient handling of permits and other red tape issues.

1.2.5 The Built Environment

In each building there are a large number of details that have an impact on energy consumption. There are a large number of energy-efficient solutions for different environments and climates that can be implemented on a large scale to make new buildings more efficient. However, the construction industry is often slow to adopt new building practices and most houses are built using traditional technologies. Transforming construction practices and supply chains for construction materials on a large scale is a tremendous change management challenge. In addition to this there is a need to refurbish existing buildings to improve energy efficiency. This means improved insulation, changes to windows with double or triple glazing, automatic control systems for light and heat, and other installations. In the field of city planning, public transportation systems need to be expanded, the frequency of departures needs to increase and the networks need to become denser, reaching out to more neighborhoods. In the United States and other parts of the world, cities need to become denser in order to make public transportation possible or in order to make it possible to expand existing public transportation systems.

1.2.6 Industrial Processes

Energy improvements in industrial processes could largely be made using existing technologies, production methods and materials. In the case of companies in energy-intensive industries it is often easy to identify potent actions that could be undertaken, but all of them are not financially justifiable for the company over the short term. In the case of companies in less energy intensive industries many small- or medium-scale improvements are possible, but improvement activities compete for attention with other investments and activities, such as business development. Large-scale information projects and support projects, especially for small- and medium-sized firms, need to be undertaken in order to make as many companies as possible involved in this type of activity.

1.2.7 Agriculture

Even the small increases in demand for grain for fuel production that have occurred during the past few years have put an upward pressure on grain prices. Maybe the yields from agriculture can still be increased somewhat, but there is also the fear that climate change, topsoil depletion and scarcity of water are issues that will demand our attention in the near future. A sustainable strategy for the future development of agriculture is needed, which includes the efficient use of fertilizers and pesticides that are made from oil. Ecological farming is an alternative with many advocates, but there are also advocates of increased intensity in farming to increase crop sizes. These arguments need to be settled in each country; a strategy needs to be made, and managed change programs need to be implemented.

1.2.8 Behavior and Work Life Change

Large-scale change needs to be managed. Behavior and work life change could only be achieved through large-scale information campaigns and training. Consumption patterns are intimately linked to economic growth, and we have substantial experience from increased consumption in the past. We have much less experience of the economic consequences of reductions in consumption, and such proposals need to be analysed both from a change management and economic perspective before they are implemented on a large scale. Changes in consumption patterns, for example from traditional light bulbs to low-energy alternatives, are less problematic than reduced consumption.

1.3   ABOUT THE BOOK

Overall, this book is written on an optimistic note. We face a tremendous challenge, but this challenge could be mastered if we approach it in a productive and structured way. Four steps are necessary to arrive at a successful outcome. We need to combine analysis, strategy development, planning and managed change in a focused way, and we need to use experiences and competence from all over the world in order to succeed. We also need more detailed management tools and organization and practices that have been tried and tested previously in government programs and other change efforts in the past, and we need to be creative and invent and deploy new management and financing solutions when they are necessary.

Part I

The Challenge

The Challenge

CHAPTER 2

Who Should Read This Book and Why?

2.1   A GUIDE FOR DECISION MAKERS

Up until now most of us have not thought about the energy that we consume. We habitually take the car when necessary, and it tends to be necessary more often than not. We are able to use the amount of electricity that we need, without second thoughts about the environment or about possible supply problems in the future. Recently many of us have become aware of climate change and the role that our patterns of energy consumption play in this development. Some of us, but still a minority, are also starting to wake up to the possible shortage of energy that is looming in the future, due to the peaks in oil, natural gas and coal production. Many people still believe that these developments will pass by, most of us barely noticing them with new fuels and energy-conserving technologies being developed at high speed as the markets signal that demand for them is increasing.

This effortless solution to the problem is not very probable and many of us will find that we, in the years to come, will have to change our lifestyles and ways of thinking in dramatic and unexpected ways. In order for this to be possible on a large scale, we will need to change many of our large-scale systems for transportation, utilities, construction, agriculture, industrial production and distribution and, not least, our lifestyles and work routines. In a sense, such large-scale change is impossible. At least it will be impossible without a rapid change in a number of large-scale systems that to a large extent determine how we as individuals lead our lives.

As individuals we are "forced" to use existing transportation systems, city plans, buildings, agricultural products and common practices in society. Even if some of us may succeed in dramatically reducing our dependence on these systems, for instance by swapping jobs and moving out into the countryside or by buying only ecological products or reducing consumption, the capacity of alternative systems is limited and someone else will buy the house and take the job that the energy-conscious left behind. If many of us manage to cut down on consumption we may find that unemployment increases and other undesirable effects on society and the economy may arise. This is not just negative and pessimistic "bullshit" from a conservative who tries to find every possible argument to maintain the status quo. No, on the contrary, I am an advocate of change, large-scale change. But for large-scale change to become possible, we cannot rely on individuals finding the way to change on their own. "Society" and "business" need to facilitate individual change by investing in the creation of new systems and new technologies, and gradually and at a relatively high speed, by replacing existing systems and technologies by new and more energy-efficient ones.

This development will require that decisions in the future are made based on a number of other criteria than those that have been used by decision makers up until now. Systems change in business and on the level of society needs to be organized in such a way that decision makers at all levels and in different parts of society use the same criteria and values when they make decisions to change systems.

This book is aimed at decision makers at all levels of business and government (local, national as well as politicians and administrators in international organizations). Some decision makers who read this will be in a position to make immediate use of the contents by starting projects or changing the direction of existing projects. Some will be able to support the idea of starting a project when it is presented. Others will be better prepared for an oncoming future situation when they may get involved in a project or be asked about their opinion on one. Only a minority of readers will be in a position to start an Apollo program or a Marshall Plan, but a large number of smaller projects in a number of different areas will eventually need coordination. This need may over time give rise to projects of a size and magnitude that may have been unexpected by the people who initiated the small projects in the first place. As we will see, the entire US space program started with a Space Task Group of 45 persons that was reluctantly formed by President Eisenhower. In due course a number of previously separate initiatives were tied together to form what we now view as a program. The same is true for the Internet, which is an amalgam of a large number of separate technologies that are necessary to send electronic messages between computers. These were developed by a number of different persons in a number of different firms, government organizations and universities and only later tied together to form a global computer network.

In the same way we will, over the next decades, see how local and national initiatives will be increasingly tied together to form new European, American and Asian systems for production, distribution and energy supply. Maybe, within two decades, entirely new companies, supply chains and industries may make up the complex systems of the global economy. This development will be driven in part by the market forces, as customers demand new products that are produced and distributed in new ways, and in part it will be driven by managed change from a high political and industrial level.

Therefore, above all, the ideas of Global Energy Transformation need soon to be adopted by many high level politicians and business leaders, and we will many times need to look away from factors and issues that have divided us in the past, in order to focus on the over-arching goal of securing the energy supply and environmental sustainability of the future. In the words of President Obama in his moving and insightful book *The Audacity of Hope*: "What's needed is a broad majority of Americans – Democrats, Republicans and independents of goodwill – who are reengaged in the project of national renewal, and who see their own self-interest as inextricably linked to the interests of others … Unless political leaders are open to new ideas and not just new packaging, we won't change enough hearts and minds to initiate a serious energy policy or tame the deficit.[1]" Global Energy Transformation argues that people and their leaders in all countries need to unite around the task of transforming energy systems. This will be a necessary change in order to secure future affluence, sustainable business practices and a basis for peace in the world. Similar to the book by President Obama, Global Energy Transformation provides no easy answers, no self-confident promises that the energy transformation will be achieved without sacrifice or partisan arguments in favor of political ideas or business practices of the past. The energy transformation could not be achieved by applying more of the ideas that created the present predicament. We, quite simply, need renewal.

2.2 ORGANIZATIONAL LEARNING – THE PROCESS THAT DEVELOPS BUSINESS PRACTICES AND SOCIETY AS A WHOLE

Developments, such as the projects, changes and technology developments mentioned above, are the fruits of processes of organization that are sometimes called "organizational learning." Organizational learning differs from individual learning in that a number of individuals in an organization or in society learn to perform complex new tasks in a new and organized manner, or change the way that a task is performed from an original situation to an improved future situation. This process is often complex and takes a long time, but through organization and planning it can be sped up and managed, so that the time and cost of learning are reduced and the quality of the result is improved. The professionals who deal with this process may be called "change managers," and more of us need to learn and apply these skills in the energy area in the years to come.

There are many aspects of organizational learning that could be described and some of them will be discussed later in this book. To some extent organizational learning is dependent on spontaneous processes and experimentation, but deliberate formalization of routines and processes is also a necessary ingredient. The market mechanism can be used and is often used in order to allocate resources to companies and organizations, and to departments and individuals in these organizations, that use them most efficiently. But, as we will see, the market is not an efficient solution to all needs of organizational change, and often, to increase speed or reduce cost (often in the form of resource waste) for society as a whole, planning and management of the learning process is necessary. Experiences from the Second World War, the Apollo program and the Marshall Plan are used in this book to argue that not only planning and management must be utilized in companies but also that these tools have been used by society and on a large scale to solve very complex problems.

As organizational learning takes place, new or changed organization structures and routines for interaction are created. NASA was formed as the time-critical Apollo program needed a formal organization and strict management in order to deliver on time. Another example of organizational learning on the level of society could be taken from the health-care system. From very rudimentary beginnings, our experiences from health care over centuries and decades have resulted in the existing organizations, where responsibilities are divided between authorities and health practitioners, family doctors and consultants, nurses and a large number of specialists in clinical chemistry, anaesthesiology and radiology, to mention only a few areas of specialization. On a higher level, the learning process has led to a division of responsibilities between universities, hospitals, pharmaceuticals companies, consulting firms and a host of other organizations. These are tied together through various patterns of work processes, rules and regulations, financial transactions, quality evaluations and red tape. All these aspects are the results of organizational learning in different forms.

From the aspect of society it is in our common interest to manage the process so that we optimize the result while we leave as little room as possible for individuals and organizations to take advantage of the situation and use these systems for their own purposes. This also calls for strong management of the learning and change process.

This book provides a framework for decision makers who want to initiate programs and projects in the field of energy transformation. It could be read as a management handbook which gives practical advice in all kinds of management situations. The aim is to provide an easy-to-understand guide, a road map for the journey into the largely uncharted territory of energy systems transformation. As we need to move from small-scale change to large-scale transformation of energy use in our society we need to make use of as much as possible of the management experience that is available to us. The book also provides background information on many subjects related to energy systems, change management and industrial processes. It is not a primer on energy technologies. It analyzes the aspects of energy systems that are relevant for change management purposes.

This means that the book should be read not only by beginners in energy matters but also by energy experts. This book does not mainly repeat information that is already known by people who have worked in energy-related sectors for years or decades. Instead, it focuses on the nontechnical aspects of energy systems transformation. Instead of analyzing the advantages or disadvantages of wind or solar power from a technical point of view, it describes the challenge of transforming energy systems. And during the process of organizational learning, even the most knowledgeable individuals need to learn new skills.

2.3 THE OUTCOME OF THE PROCESS: THE GLOBAL EMBRACE AND CONTINUED PROSPERITY

Some authors have raised alarm that the peaks in oil, gas and coal production may lead to resource wars that may even be the end of civilization as we know it. While this is certainly possible, good preparations and an organized approach to the issues will reduce the probability of such a war.

The basis of our prosperity, the increasingly global economy and economic growth, derives its strength only from the faith that people all over the world have – that this system will be able to create virtuous economic cycles in the future as well as it has done in the past. In a situation of possible soaring energy prices, this system will come under great pressure. In such a situation, decision makers at the highest levels of the large countries of the world will have a few choices, most of them falling under the categories of "cooperation" or "conflict." If these decision makers share the view that it will be possible for the world to live happily together in a future world fueled by renewable fuels and in which much less energy than today could be used, they may collectively embark on this type of path and cooperate to solve the energy problems and create a better future. In the absence of a vision of how cooperation could be organized, decision makers may resort to violence and conflict, instead of creative cooperation.

In a future of increasing energy prices there will be tremendous pressure on the global economy, and the relations between nations that are all heavily dependent on scarce energy resources will be put under pressure as a consequence of this. The faith in continued economic growth may be shattered. This in itself would put economic growth at peril. Global economic systems and national currencies will have to be renegotiated and, new bases for economic stability and financial exchange will have to be established. This process will require global cooperation between nations under very difficult circumstances, as all nations have the incentive to try to achieve agreements to their own advantage. The force that is likely to balance the drive to suboptimize in this way will be the global interdependence between nations when all nations today, in different ways, are dependent on others. No country would, at short notice, be able to function on its own.

If we, on a global basis, run this process in a cooperative manner and refrain from large-scale conflict and resource wars, we may experience a prolonged period of international cooperation and restructuring of the global economy, which we may coin "The Global Embrace." In some ways it may be similar to the Cold War, as an example of a prolonged period of relative peace under difficult circumstances, all major participants realizing that the alternative would be worse.

The contents of this book form a road map for the decision makers at all levels, who need to understand the nature of the challenge of energy transformation in order to lead the way through potentially difficult times.

CHAPTER 3

The Sustainability Challenge

We are constantly bombarded with reports about pollution, climate change, the imminent production peak for oil and other pressing issues that demand us to develop new technologies, products and services, business practices and lifestyles. It is becoming increasingly clear that a substantial share of the climate change that we can observe has been caused by human activities. It is also now clear that an increased focus on sustainability will open up substantial business opportunities for the companies that take part in the change and develop the technologies, products and services that will be needed in order to facilitate a transformation to sustainable energy systems. We have also been informed by the scientific community that the alternative to major change may be further large-scale changes in the climate, which may alter the possibilities for life on this planet for the longer term.

Many of us now agree that it is time to introduce the issue of sustainability much more strongly into the business community and also into the minds of consumers, whom we expect to buy the more energy-efficient and less polluting products of the future. In many parts of the world, this realization has come about recently, with an increased interest in energy-efficient technologies and other business opportunities from investors and companies as a consequence.

The environmental reasons and the reasons that are related to oil production may in themselves be strong enough to justify a strong wave of investments in the areas just mentioned. However, those are not the only ones. Contrary to the beliefs of most people, the lines of business development that have been pursued in the past cannot, for very clear business reasons, be continued forever. Economic growth along the paths followed by our society for centuries will at some point generate diminishing returns. This is for a very simple reason. Economic growth in our system of finance, currencies, banks and companies requires continuous productivity improvement in order to continue. In a previous book by the same author, it has been argued that the opportunity to continue to reduce the cost of present operations in the development, marketing and production of existing goods and services cannot continue forever. Thus, we constantly need new areas to invest in, and areas in which future efficiency improvement can take place.[1]

Frankly, we do not know whether this will always be true, but in this case, as the opportunities for further business development in the old areas of strategic dominance and operational excellence are starting to show diminishing returns, the new area of sustainability, where energy efficiency has a very important role, will prove to be the next wave of business development. Simply put, as a society we now need to clean up the mess that our efforts at global business and operational efficiency have put us in. This needs to be done rather quickly so as not to get burned by the increasingly hot climate that we seem to have caused. We also need to increase the supply of renewable energy and, in particular, of renewable fuels for transportation, since the oil supply is no longer growing and we need ever-increasing amounts of energy in order to maintain economic growth.

Productivity improvement is a prerequisite for economic growth. Productivity growth comes from the gradual cost reduction that has been a consequence of efforts toward market dominance, operations improvement and other business practices of the past and present. Over the years, however, we have been lucky enough to enjoy the benefits of cheap energy. Oil has been one of the primary sources of energy to fuel our world, and it has been cheap. The increased use of energy during the previous century strongly correlates with economic growth.[2] For a long time the price of oil has hovered around 20 dollars per barrel. Other energy sources have also been relatively cheap and abundant. Now, as oil and other sources of energy increase in price, the opportunity for alternative energy sources to be introduced on the market is opening up. At a price of oil above 50 dollars per barrel, the financial prospects for the development of alternatives are rapidly improving. Since we have no such sources that are readily available on a large scale, we will need to develop some of the sustainable technologies, or adapt existing technologies to new fuels and raw materials. In some cases we merely need to expand production capacity for technologies, products and services from the current very low level of production. This development will require large investments, but it will also create substantial, and many times global, business opportunities.

While political and business leaders are just starting to find their feet in a terrain where energy efficiency takes an increasingly important position, I am now raising a call for change management and leadership at three levels in society and economy. We need leadership at the following levels:

– Transformation of national energy systems (the program level in this book).

– Transformation of complex global sector energy systems, involving a number of different companies and other stakeholders (the stream level in this book).
– Transformation of energy technologies in companies and particular projects. This includes a transformation of the ways that companies use energy, and how energy use could be transformed through the development of new technologies, products and services (the project level in this book).

While the debate up until now has been focused on market-driven change that will be directed by the market forces left on their own, I argue that there are four steps that will definitely become necessary in order to drive focused and speedy change:

1. **Analysis** will be needed at different levels of society and business. First, an overall map of the territory needs to be drawn and maps also need to be made for each sector. This work is started in this book. The analyses need to identify the areas in which change can be handled by market forces and the areas where investments will not be financially justifiable and where government financing or financing from sources other than financial markets will be needed.
2. Based on the analysis a **strategy** will have to be developed. Some of the key aspects of this strategy are described in this book. More issues will probably arise during a more thorough analysis, but we need to start with what we know already.
3. **Plans** for the change need to be drawn up and a realistic time frame for change in each area needs to be determined.
4. In order to drive change at the pace and in the direction that we decide, **change needs to be managed**. This will require a change program consisting of subprograms, which I call "streams" (one for each sector discussed in Part 3), and projects for the technology development, commercialization, expansion of production resources and market penetration of the different initiatives in the program.

In order to manage change, our society needs business leaders and political leaders that are able to lead the way into the uncharted territory of large-scale business and social change. While it can be argued that the emerging field of business development for sustainability is a completely new area for transformation and business endeavors, I argue that the world in general, and the United States in particular, has several past experiences that can be drawn upon in the oncoming change process. Past American successes, such as turning the US industry into war production, arming the Allied forces in order to win the Second World War, in only three and a half years, is a remarkable feat that we can learn from in the future turnaround situation, which, as some experts argue, has the urgency and the level of investment and effort needed in common with the Second World War effort.

Another American success story, which has contributed substantially both to business and economic development in the past 50 years, is the American space program. Contrary to experience from the Second World War, the space program has been run in parallel with other economic activities, but business development and economic growth have benefited tremendously from this long wave of investments.[3] We will discuss the Apollo program, and some of the lessons learned from this. We will also describe some aspects of the Marshall Plan, and discuss some experiences from this effort, which represents a post-war aid program, for the reconstruction of European industry, in which aid was focused on particular business sectors in the economic reconstruction of Europe.

The sooner we analyze the opportunity of running the transition to sustainable energy sources as global, regional and national programs, sector-based streams and company projects, we can start to understand the need for planning and management of the program. Such a program will need to be managed on a national basis similar to the way that the space program has been managed in the United States. Probably a program will also be needed on the global or transnational level. At these high levels, we will need to make decisions about which sustainable technologies to choose and develop, we need to set overall goals for energy consumption and CO_2 emissions and we need overall national and company-wide management processes to achieve these goals. We also need widespread information and training of managers and other participants in change projects in methods, procedures and work practices that need to be developed for this purpose.

Overall, the energy transition may not necessarily be a problem, but an opportunity. It represents an opportunity to renew energy production and distribution, production machinery and industrial installations, and transportation technologies and systems. This, in itself, represents an economic opportunity to initiate a very positive development, at a time when there are worries for business and economic development in the near future. The energy transformation does not represent a hopeless situation, but the opposite.

As a society, we are already in possession of many of the strategic tools and change management tools and principles that we need to employ during the transition. As we will see below, we also have access to a number of technologies and products, in different stages of development, which will represent some of the solutions that we need in order to improve energy efficiency. In this book the overall principles for change are discussed, and a toolbox and management processes are developed that can be used at all levels of society and in business, in order to manage the change. An idea for an overall management structure is also introduced, which can be used in order to organize and run the program.

There are a number of key arguments that are recurrent in this book:

- The solution is not only technical, financial, or a matter of more conscious consumption, or of management. The solution does not rest with any one particular group, or stratum, of society. The solution will be a combination of a number of different aspects of a solution, and it will need to be a truly collective effort. While the technical aspects ("what" to change) have been widely debated, much less focus has been given to the business and change management aspects of the change ("how" to run the change). The latter will be discussed in-depth in this book.
- The markets of the world will not be able to rapidly create these complex solutions, without the help of planning on the overall level of society. We need to apply the four steps necessary for change that are introduced in this book. Otherwise, we will not be able to stick to tight time schedules and ensure precision of results.
- Planning of transformation needs to be firmly based on an understanding of the limited resources that we need to use in order to facilitate the change. We need to optimize the plan for change, based on the idea that we have to use the limited resources as efficiently as possible. The most important limited resources are as follows:
  - time;
  - money;
  - raw materials for renewable energy;
  - land, for the cultivation of energy crops, construction of solar photovoltaic plants, wind farms, etc.; and
  - competence in the relevant technical, financial and change management–related areas.

In addition to these, the book also aims at achieving the following:

- To put the need for sustainability in the perspective of recent management history and argue that business needs a new area for investment and growth and that this area could be sustainability.
- To outline some of the possible challenges and developments of a managed change effort and describe the opportunities for economic growth that are embedded within them.

- In order for structured change to happen fast, change management is needed at several levels. In this case change management, probably, will mean that different players in society make plans and take action in a continuous dialogue with other players and also that each player is willing to revise plans and change the course of development depending on the outcome of this dialogue. This book provides the tools for strategic planning, experimentation and also for managing the program at all levels.
- To consider what strategic alternatives are open to companies that are interested in pursuing sustainable ways both to increase operational efficiency and to innovate new products, which may appeal to consumers and businesses that are ready to work with sustainable business approaches.
- To examine both risks and new opportunities currently arising due to changes in the environment and the global marketplace.
- To look at various possible road maps for organizational development where leaders have made a commitment to working intensively in the crucial decades ahead, in which scientists say that we must make changes around the world in order to ensure that our children and grandchildren will have an acceptable quality of life.

Box 3.1 We Need a High-Speed Transformation Program

An increasing number of experts agree that our planet is heading for dramatic changes in climate, and that we only have a few years to change our energy systems on a large scale. If we do not manage to accomplish this, we run the risk of climate and ecological systems being put permanently out of balance. According to these experts, we need to, dramatically and rapidly, reduce CO_2 emissions.

- "The global society has nine to ten years to come to a binding agreement and begin dramatically cutting the emission of green house gases if we are to avoid disaster." James Hansson, NASA's chief scientist.
- "I am more afraid of climate change than the threat of terrorism." Hans Blixt, United Nations Commissioner who investigated Iraqi military arsenals looking for "weapons of mass destruction" under the close scrutiny of Saddam Hussein.
- "The [IPCC's] Fourth Assessment Report is a milestone in our scientific knowledge about climate change and the grave threats global warming poses to the planet. The report's findings amount to a stark warning that the world must act fast to slash greenhouse

gas emissions if we are to prevent climate change from reaching devastating levels. The good news is that it also shows that deep emission cuts are both technologically feasible and economically affordable." EU Environmental Commissioner Stavros Dimas.
- Dr Rajendra Pachauri, Chairman, United Nations, Intergovernmental Panel on Climate Change (IPCC), stated in his speech of acceptance for the Nobel Peace Prize: "At a fundamental level the world has to create knowledge and practice on a path of development which is not (natural) resource degrading and carbon intensive. Human ingenuity and strength are capable of meeting this challenge.... We need to commit ourselves to that path before it is too late."

In order to achieve this, an international agreement, the Kyoto Protocol, has been signed. Through this protocol, nations of the world have agreed to reduce CO_2 emissions within the next decades. The exact levels at which CO_2 emissions will be held are not yet decided. Over the next 2 years, negotiators from each country will work to create a "deal" which is robust enough to hold up under the final and decisive round of talks to be held in December 2009 at Copenhagen. Here, a set of new policies and regulations limiting CO_2 emissions is scheduled to be signed by the countries of the world.

The Kyoto Protocol is an important step in itself, but, according to experts we need to reduce emissions faster than the goals of the Kyoto Protocol demand. It is even possible that the global business community does not meet the goals set up in the protocol.

Market-based solutions, such as the Kyoto Protocol, are probably too slow, and unfocused, to facilitate change at the rate that we now need. We probably need to drive both technology development and the implementation of technology at a faster pace. This is both because of climate change and because of the peak oil issue, which threatens to increase the price of energy to previously unseen levels.

The key elements of EU's Environmental Action Plan 2007 are as follows:

- a commitment to reduce the 1990 levels of greenhouse gas emissions by at least 20 percent by 2020, which will be strengthened to 30 percent reduction in the context of a fair global agreement;
- a firm target to increase the use of renewable energy to 20 percent by 2020;
- a broad range of measures to improve energy efficiency by 20 percent by 2020;

- further evolution and strengthening of the EU's emissions trading scheme;
- an ambitious limit to reduce CO_2 emissions from cars;
- a framework for introducing carbon capture and storage (CCS) in power production; and
- development of an effective adaptation strategy.

One key argument of this book is that the market-based approach of the Kyoto Protocol and the expectation regarding efforts to voluntarily conserve energy and reduce emissions will not be sufficient. We need to complement market-based efforts with planned change. The country that has the strongest experience of large-scale, planned, business transformation projects is, arguably, the United States. The efforts, previously mentioned, such as arming the Allied forces for the Second World War, running the US space program, and contributing to the rebuilding of Europe after the Second World War, through the Marshall Plan, and a number of other such efforts, represent examples that we can, and need to, learn from.

If we apply the lessons from previous large-scale industrial change efforts, develop a strategy based on the best facts and figures that we have at our disposal, and run a project that combines planned efforts with market mechanisms for technology, product and business development and economic growth, we can use this situation to our advantage. As a society, we may even come out stronger in the end, than we were at the beginning.

There is a widespread belief in society that market-based change, by necessity, is to be preferred in all situations and that planned efforts, wholly or partly, financed by government, are inefficient, and, potentially, lead to economic decline. This is not the case. The experiences from former Eastern European, Chinese, and Cuban socialist economies are not relevant experiences to use as guidelines in this case. In these economies, public ownership of the factors of production and the lack of incentives to improve society and its systems were important aspects that will never be part of any planned program in Western economies. The relevant examples for large-scale planned change, combining market mechanisms and planned effort, are the type of examples already mentioned. In these cases, as we will see in the case studies, planning, at the level of society, adds clarity of purpose, speed and determination. Market-based change may provide consumers with the exact type of automobile that they want, but it is a relatively slow and unpredictable way to transform society or to achieve other highly specific results. This argument will be developed further later in the book.

Four Scenarios for the Future

In the current literature a wide range of visions of the future are emerging. There are people who envision a future that will unfold as a continuation of past trends of increasing prosperity through technical and economic development. This may still be the most widespread picture of the future. I call this scenario "business as usual." There are also experts in a wide range of fields, such as climate, the environment and peak oil, who envision futures that are substantially different from the present and where forces that have had a limited role historically take center stage and transform our environment and society in ways that most of us do not expect or wish for.

Depending on the magnitude of change that different analysts foresee, the dystopic scenarios could be based on a low level of change, where society is able to handle disruptions through a strategy of "muddling through." In case climate change could be reversed within the next few decades and if the price of petroleum would go through low-to-moderate increases, this scenario may materialize. If, on the other hand, we face a period of major disruptions in climate or severe shocks in oil prices and in the prices of other types of energy, raw materials and manufactured goods, we may encounter a period of severe ecological and economic turmoil. This scenario is called "environmental and economic disaster."

In this book, this dystopic view of the future is positioned against a future that I call the "ecological dream" (Figure 4.1). This is based on the hope that humanity collectively manages to master and thrive on the challenges that we have set up for us and that we also manage to build a completely new type of society by making key changes in a number of technological, economic and social systems. This may be seen as naïve by readers who are already knowledgeable about the immense challenges that lie before us. It may be possible to solve many of the problems that we have created, but it is unwise to underestimate the challenges. It is also unwise at this point to give up hope for a better future for our children and coming generations. During the writing of this book, I have also become increasingly aware of the amazing management achievements of the United States in the past. If we, on a global scale, manage to repeat some of the successes of the past, and successfully take on the transformation challenge within an entirely

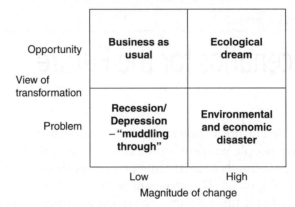

	Low	High
Opportunity	**Business as usual**	**Ecological dream**
View of transformation		
Problem	**Recession/ Depression – "muddling through"**	**Environmental and economic disaster**

Magnitude of change

Figure 4.1 The four different scenarios found in the literature about the future

new area of endeavor, we may be able to reverse climate trends and preempt energy shortfalls. This blissful future is, therefore, presented as one that is still available for us to strive for.

Below, the essential elements of each of the scenarios for the future are recounted; and the argument that we probably have to expect a high magnitude of change, with the possibility of maintaining a positive view of the future, is put forward. The benefits of an energy transformation effort may be stronger than what any of us currently expect.

4.1   THE FIRST SCENARIO: BUSINESS AS USUAL

Based on decades of warnings, which many believe to be false alarms, many people fear that the natural disasters and extreme weather experienced over the past few years, in the form of hurricanes, forest fires, droughts, extreme rains and tsunamis, are unfortunate, but temporary aberrations, caused more by chance, than by permanent or semipermanent changes in climate. People who live in areas used to a colder climate may welcome the prospect of having Mediterranean temperatures and look forward to increasing tourism in the summer. It may be tempting to see the consequences of global warming as a zero-sum game, in which some countries win short term because of increased tourism and other advantages, while other countries receive more rain and thus may get slightly better conditions for agriculture.

Overall, in this vision of the future, past patterns of economic growth may be sustained. Disasters and the reallocation of industries and opportunities may even lead to an increasing need for investments, which may drive economic growth in the longer term.

In this vision of the future, the peak in oil production that seems to be on the way should not cause a problem; it could instead present us with an opportunity. We have all heard of a number of new energy technologies, based on renewable fuels or "free energy" that are on the drawing boards of scientists and in universities and technology development companies all over the world. The more optimistic of us may believe that these technologies can be rolled out globally during the course of a few years and that the implementation of these technologies represents a huge opportunity to fuel economic growth once more. In this scenario, the belief is that market forces will allocate money and other resources to the most promising technologies at the pace that will be needed, in order to solve the problem. There will be no need for planning at the societal level. Let the market solve the problem and create the technologies and resources in the way that it has always done in the past!

4.1.1   Assumption about Economic and Social Disruptions

In this scenario of low-level change and a positive view of the future, this could be done without a "big fuzz." The change may actually go unnoticed by most people, who will experience only how a number of large power plants based on renewable fuels get built during the coming decades and how an increasing number of cars and trucks become equipped with hybrid engines, fuel cells, electricity-based systems and other renewable fuel systems.

According to the proponents of this view, there will be no disruptions; most of us may not even recognize the change as such, only noticing a number of new technologies that gradually solve what was never really a problem at all.

4.1.2   Assumption about Leadership

In this scenario, there is no particular urgency or need for mobilization of resources or management of a "change program." Everything will be taken care of by the market. The "wisdom" of the market with mechanisms to immediately divert resources to where they are needed, using its "invisible hand" to facilitate the process, is not just immense. The market is amazing, immediate and almost completely flawless in its wisdom.

According to this view, the reason for the market not already diverting substantial resources to sustainable technologies is that it has not been necessary. As soon as it becomes necessary, the market will make the necessary means available at once. People and companies will know what to invest in, and the market will allocate the resources; it is as simple as that.

Even if proponents may concede that the market is not exactly flawless and immediate, there is no better alternative. Planning is not an option in a market economy. Period. This has been proven by the dismal failure of the Soviet economy and other planned economies. The less planning there is in a market, the better the resource allocation works and the higher is the growth. It is as simple as that.

4.1.3 Example of a Proponent of This Scenario

Most business strategists and economists see the future as being no different from the past in terms of economic and business opportunity and growth prospects. The authors of the book *Green to Gold*, Daniel Etsy and Andrew S. Winston, seem to be proponents of this view.

4.2 THE SECOND SCENARIO: RECESSION/ DEPRESSION – MUDDLING THROUGH

By now, many people would have heard of the tremendous cost of the environmental disasters that we have already experienced and they may even assign substantial credibility to the forecasts that depict such events as becoming increasingly numerous in the future. The same may be thought about forecasts of increasing prices of petroleum and other energy sources. If we have to spend a lot of money just in order to rebuild things that we have already built once, and if we need to spend increasing amounts of money on energy that used to be very cheap, which may prevent us from increasing our spending on things that could improve our lives or society, people may conclude that there may not be much room left for economic growth.

The more realistically minded of us may also acknowledge that it usually takes years or decades for technologies to be developed from idea to prototype and into marketable products. From the launch of a product in the market, when it is usually quite expensive, to the point of mass marketing of inexpensive varieties of a product, it often takes decades.

Based on experiences from the past oil crises and other temporary disruptions, many people may conclude that it is still a realistic option to get through a period of difficulties by muddling through. We may have to drive less often to the supermarket, experience a shortage of exotic fruits and other products from afar and maybe people with less well-paying jobs will have to travel by bus or train to work, while the more affluent will be able to maintain much of their lifestyles.

Overall, disruptions and shortages may lead to a period of recession or depression, which may last for a few years or a decade until we have aligned our technological and economic systems to the new circumstances. From this point onward, economic growth may continue indefinitely, following familiar patterns of the past.

4.2.1   Assumption about Economic and Social Disruptions

According to this view there will be change, but the exact nature of change is difficult to determine in advance. We will learn about change needs and possible solutions as we proceed and solve problems as they arise. It is not much use trying to foresee problems at this stage because we are either too late or we have too little knowledge. Based on very impressive business achievements in the past, we will probably be able to handle problems and muddle through.

There may be social disruptions and sometimes abrupt changes in the direction of economic development and other factors in society, but any problems will, unfortunately, primarily affect the poor and less educated. Educated people with a good position in society will be important resources throughout the muddling-through process and there will always exist opportunities to prosper, even in the hardest of times. People struck it rich during The Great Depression, and in the end it is only up to our own imagination and drive how well we do in business and society.

4.2.2   Assumption about Leadership

Leadership will be seized by people who happen to have the right resources for the moment. The right people will come to the fore and lead. This will create opportunities for companies that offer the necessary technologies and services for the change. "Creative destruction" will befall those who do not manage to change or who happen to be in the wrong place at the wrong time.

4.2.3   Proponents of This Scenario

Stephen Leeb, in his book *The Coming Economic Collapse – How You Can Thrive When Oil Costs $200 a Barrel*, argues that business-minded people should plan their investments today so that they prosper in the event the price of oil is 200 dollars per barrel. He also tries to identify other business strategies that could create wealth and prosperity in the possible hard times that lie ahead of some of us.

4.3   THE THIRD SCENARIO: ENVIRONMENTAL AND ECONOMIC DISASTER

The more pessimistic readers may fear that society will face disruptions that are so severe as to cause environmental and economic disasters of a magnitude that we can hardly even fathom. Society may also suffer prolonged shortages of oil and other energy sources, which will disrupt production and distribution of goods and food and, which may, possibly, also lead to starvation, sickness and death of millions, even in developed parts of the world.

This situation may unfold over decades and problems that initially seem possible to handle and a development that may look reversible, may, in the eyes of the more pessimistic analysts and members of the general public, turn out to be irreversible and ultimately fatal for many.

Even if this development, at some point, comes to an end, the more pessimistic may draw parallels to the Middle Ages, or Dark Ages, in Europe that started just before 500 AD and lasted for a thousand years. The cause of the initial economic decline and social disruptions was the intrusion of Asian peoples, such as the Huns, into Europe. Even if there were prosperous periods when, for instance, the great cathedrals were built, the real economic upturn did not start until the Renaissance, when a number of new ideas took hold and European society started to open up.

4.3.1   Assumption about Economic and Social Disruptions

According to this view, we will see a dramatic decline in affluence in society and a prolonged period of depression, shortages and disease. Economic change will be tremendous and uncontrollable shifts in power structures will occur. This may even be the end of nation-states and modern society as we know it.

4.3.2   Assumption about Leadership

Beyond some point in the development, the proponents of this scenario seem to believe that it will not only be difficult but also impossible to exercise leadership across states and other geographically large territories. Society will break down and we may see a repetition of the experiences of the Middle Ages, when villages closed in on themselves and trade and ideas' exchange with the rest of the world broke off.

4.3.3   Example of a Proponent of This Scenario

James Howard Kunstler is, probably, the most prominent proponent of this scenario. In his book *The Long Emergency – Surviving the Converging Catastrophes of the Twenty-First Century*, he paints a very dystopic picture of the future, and apparently, while driving his car in the New England countryside, he can already picture himself in his dystopic future setting.

4.4 THE FOURTH SCENARIO: THE ECOLOGICAL DREAM ACHIEVED THROUGH THE GLOBAL EMBRACE

Still assuming that we will have to face a number of severe problems, the scenario of "the ecological dream" suggests that we will be able to solve these problems, avoid the most severe disruptions and actually rebuild technology and economic and social systems on a massive scale. In order to achieve this we will need to go through the period of "the global embrace", during which we will have to reorganize society toward more energy-efficient systems, business practices and lifestyles. We will succeed in transforming energy systems and society at the required rate in order to roughly maintain economic and social stability and technological and economic development.

This scenario suggests, for instance, that we will be able to reduce our dependence on fossil fuels at approximately the required rate in order to avoid exorbitant increases in energy prices. It also suggests that we will be able to reduce CO_2 emissions and emissions of other greenhouse gases rapidly enough to stop global warming at the present level.

This book argues in favor of this scenario. The main argument is that the only tool by which we could hope to achieve this is a planned and managed process that is run in parallel at all levels, and in all areas, of society at once. The process resembles, in its management and planning mechanisms, more the planned development and achievements of the US space program, than the planned economies of the former Eastern Bloc. We do not want to end up in a situation where we have too little time left to avoid disaster and need to run a total and immediate turnaround of the global economy, to form a peacetime equivalent of a war economy, that is totally focused, from a development and resource perspective, on one single goal, though we can also learn from the US World War II effort. Instead, we want to transform the economy and energy systems in parallel with other economic and social activities and come out of this process as a stronger and more affluent society with better opportunities for all.

4.4.1 Assumption about Economic and Social Disruptions

In this scenario, the assumption is that there may be severe strain on economic, technical and social systems, but this strain can be foreseen, planned for and dealt with in a creative fashion. It is assumed that technologies can be developed and systems can be changed at the required rate in order to maintain stability and order. It is also assumed that it will be possible to inform and train people, managers, politicians and other key participants in the change process, so that we all will be able to participate, and act, in a sensible and productive way.

4.4.2 Assumption about Leadership

In this scenario, leadership and planning will play a very important and fundamental role. It is this leadership and change management that is the topic of this book.

4.4.3 Example of Earlier Proponents of This Scenario

Richard Heinberg and Colin Campbell have made very important contributions to this line of thinking through the book *The Oil Depletion Protocol*. Before this, Richard Heinberg wrote *The Party's Over* and *Power Down*, which started to chart some of the opportunities for energy transition. *The Oil Depletion Protocol* represents the most structured approach. In this book the authors present the idea that municipalities, regions and nations need to sign up for an "Oil Depletion Protocol" and agree to a voluntary plan for the reduction of petroleum consumption. The present book develops, in more detail, the tools for setting more elaborate savings goals for all energy sources.

Box 4.1 Climate Change, Peak Oil and How These Developments Could Influence Economic Activity

The average increase in global temperatures of 1° that we have experienced in the past decades will cause the transfer of some types of economic activities between countries. Areas that have previously been arid may in the near future become desert. In many parts of the American West it has, due to drought, become increasingly difficult to raise cattle. Climate experts forecast that one-third of the world's population will be without natural access to fresh water in only a few decades. This causes a substantial need for investment in a large number of different areas in order to both maintain conditions for life in these areas and avoid social unrest and large-scale migrations. We are approaching a point where it is becoming increasingly obvious that we will not be able to avoid the cost of remedying these problems. We can only decide to mitigate in a planned and structured way, before major problems arise or to wait and take action in the midst of possible turmoil.

In the past decades we have also increasingly experienced extreme weather, fires and drought. It is probable that such events will occur with increasing frequency in the future and that they will cause temporary or permanent damage to economic systems.

However, despite the magnitude of such occurrences, it is still not clear how transformation should be financed or how the economic signals to markets from such events could ever create a strong financial incentive for major stakeholders to take action. The occurrences happen randomly and it is seldom predictable which stakeholders will suffer the largest damages in any given year in the future. It has been pointed out by peak oil experts that oil price volatility may make it difficult to attract investments in renewable energy sources. In the case of the cost of climate change and extreme weather the situation is not just volatile. To a substantial extent it is even impossible to forecast in detail and on a stakeholder-by-stakeholder basis who will be the next victim. This creates no basis for market-based action. For this reason planned action on the part of society may be the only alternative to achieve the level of investment needed to stabilize climate and reduce the frequency of these disasters.

In the case of the production peak for oil, this has very clear financial consequences. Increasing prices and possible shortages will affect companies in a negative way. However, the signals to invest in renewable energy systems and other green technologies will come too late. Since it takes years to develop the technologies and production resources and increase the market penetration of new technologies, by the time we start to notice the effects of reduced production levels it will be too late for us to take planned action. We need a program of managed change that is put in place before production volumes decline and prices start to increase.

4.5   HOW TO AVOID DISASTER AND SUCCESSFULLY ARRIVE IN A PROSPEROUS FUTURE

When we, as a society, contemplate the prospect of changing society into a more sustainable future and lifestyle, it is easy to underestimate the challenge. At present the public debate shows signs of this. Despite the fact that the former Vice President of the United States Al Gore has received large amounts of well-earned publicity and an Academy Award for his book and film *An Inconvenient Truth*, and a Nobel Prize for his work to increase awareness of the challenges ahead of us, little urgency among policy makers, companies and the general public could be noticed. Few decision makers are willing to take decisive measures to meet the high-level goals that were presented by Mr Gore. While this could, for the short term, be comforting, it could be feared that a focus on climate and energy issues, which would inevitably materialize during the next few years, turn public sentiment dramatically in the opposite direction, toward urgency, and, possibly, even panic.

Planning the transformation ahead of events will prove justified. The ability to present not only good intentions but also a detailed plan (to present to those who care to investigate the basis for any public announcements about the preparedness for energy transformation) will be useful in jump-starting the process, and in mobilizing the attention and energy of important stakeholders in society.

4.6   WE HAVE BARELY STARTED THE TRANSFORMATION

There is definitely a risk at this time that people may interpret early success stories of promising new technologies for car engines, or renewable electricity, as if these were signs that the environmental and emissions problems were about to be solved. This is, however, not the case. We have not even begun to solve the problem or take advantage of the opportunities that the transformation is presenting us with.

While some visionary companies have started to transform their operations to CO_2-neutral mode, these companies would have, effectively, used up a portion of the easily accessible renewable energy sources, such as ethanol, surplus heat from industries and electricity from renewable sources. The sources that we have started to tap, which largely come from relatively small-scale production facilities, or which are produced as secondary products of production processes, set up for other purposes, are not inexhaustible, and we cannot continue to rely on sources that are available on a small or limited scale for the long-term transformation of our large-scale energy systems.

Box 4.2 Large-scale energy systems

In our society, we have become accustomed to using large volumes of energy every day to get by. The large-scale use of energy for transportation, electricity, heating, industrial processes and petroleum-based fertilizers and pesticides has made possible the creation of industrial agriculture. Through this system, a small fraction of the population produces and distributes food and products for the rest of us, so that we can spend time doing other things. This has never before been possible in society. Historically, most people had to be involved in food production on a daily basis. We now use fuel for transportation, heating and industrial processes, so that we can distribute goods and services across markets.

In a certain sense we are like fish that are unaware of the water that surrounds it. Our energy systems and our economic systems are often taken for granted and they often go largely unnoticed until they malfunction. This may happen when someone becomes unemployed, or when we experience a brownout in the middle of winter. Suddenly we become aware of our dependence on these systems.

As an example, the United States consumes about a quarter of the world's energy per year. In total the energy consumption of the United States amounts to 99.3 exajoule per year. The energy consumption of the United States is dominated by oil and gas. The energy mix of other countries is usually dominated by oil for transportation and different fuels for the production of electricity.

The global oil production is 1000 barrels per second, or 85 million barrels per day. The global energy consumption in total amounts to 204 million oil barrel equivalents per day. This amounts to a global average of some 7 liters of raw oil per person per day. Considering that energy consumption is higher in the developed world, than in less affluent countries, we may realize that we have a substantial task ahead of us. In order to understand the size of the transition task, we could compare with grain-based ethanol production. If the United States were to replace its use of oil by using grain-based ethanol instead, we would need to use all the land that is available in the country to grow grain for the purpose. There would be no land left to build cities, and we could use none of the grain as food. To quote Richard Heinberg: "There would be no land left over even to house the American population, let alone feed it."[1]

In order to understand the transformation challenge of the electricity system, we could use another example. The European Wind Energy Association has set it up as a realistic goal to produce 12 percent of the global electricity production from wind by 2020. This will require 1260 new GW of wind generating capacity.[2] This would mean 630,000 new 2-MW wind turbines that need to be installed by that date. The global capacity for the production of new wind turbines in 2002 was 7 GW per year. This has now increased to 20 GW per year in 2007. In order to produce the necessary turbines, we will first need to substantially expand production capacity. At the current rate of growth we will arrive at some 5 percent wind power globally by 2020 as opposed to the 12 percent that was seen as realistic.

Furthermore, in addition to the 12 percent electricity generated from wind, we would need more capacity from other renewable sources. This will require the development of other renewable technologies for electricity generation to the point where they become competitive from a cost perspective. Currently, wind power, together with cogeneration using biofuels, is the only renewable form of electricity that has reached this position, and that can be applied on a large scale.

These examples indicate the magnitude of the task of transforming energy systems. However, they nowhere nearly indicate the overall complexity of the task. In order to gradually reduce our dependence on oil for transportation, a transition for which we do not have a readily available alternative, we may need to use large volumes of electricity to fuel cars. This would increase the challenge of the expansion of electricity production even more. If the total volume of petroleum-based fuels in the United States should be replaced by electricity, the equivalent of 500 new nuclear reactors would have to be built. Currently, there are some 450 reactors in use globally.[3]

We need alternatives that can be applied on a large scale in the short term, and we need to develop more of these for the longer term.

In addition to this, some of the short-term solutions will be viable only on a limited basis for the longer term. For example, the grain surplus of the twentieth century, which could be used for the production of ethanol that could be used as a fuel for cars, has decreased and turned into a deficit in the early years of the twenty-first century.[4] There is also a limit to the amount of land that we can use in order to grow raw materials for ethanol and other fuels. In the long run we cannot rely on using increasing amounts of valuable farmland for fuel. Instead, we will need to develop renewable fuels that do not compete with food production.

4.6.1 Peak Oil and High Noon

One of the reasons why we need to cut down on our use of energy very rapidly is that an increasing number of energy experts warn that the global peak in oil production is imminent and that it may, in fact, already have happened. Similarly, production of natural gas in the United States has reached its peak.[5] A reduced supply of petroleum and natural gas will reduce our ability to produce the current volumes of goods and services and ultimately limit our prospects of economic growth. This conclusion that the global peaks are approaching is based on a number of different types of analyses:

- Some petroleum geologists have for a long time warned that the peak in oil production would occur sooner than the oil companies have foreseen. Retired oil geologist Kenneth Deffeyes repeated the calculations of the first alarmist of the 1970s, M. King Hubbert, and found that the global peak would likely occur some time before 2009.[6]
- The amount of new oil wells found has declined decade by decade since the 1960s. The amount of new oil found every year now is less than one-third of annual consumption, so that the known resources decline year by year.
- Out of the 65 most important oil-producing countries, 54 have already reached their production peaks. The world's largest oil producer, Saudi Arabia, has possibly yet to reach its peak. The United States did reach its peak in production in 1970 (as was predicted by Hubbert in the 1960s).[7]
- Even the ability of Saudi Arabia to expand production has been called into question. In his book *Twilight in the Desert*, Mathew Simmons makes a detailed review of all Saudi oil fields and of the technologies used in order to maintain a high level of production, and concludes that the methods used (e.g., horizontal drilling) would only be justified if many of the most productive oil fields were approaching their peaks. He, therefore, questions the assertion that for a long time Saudi Arabia will be able to, as a "swing producer," make up for the shortfall caused by the production peaks of most of the other large oil exporters in the world.
- Untapped resources are primarily located in deepwater locations, in arctic areas or in other remote places where it is both difficult and costly to produce oil.
- New "giant" fields that are expected to come on line during the next decade are located in the Caucasus and the Gulf of Mexico. None of these fields is expected to produce more than 250,000 barrels per day at their peak and this peak level will last for less than a decade. With a daily global production of 85 million barrels per day in 2007, we need to find a large number of such new giant oil fields, in order to push peak oil into the future. The single largest oil field in the world is the Ghawar in Saudi Arabia, which stands for 8 percent of global and 60 percent of Saudi Arabian production. There is no other oil field that comes even close to matching this size.[8]
- There exist large reserves of oil, which are tied up in oil shale and oil sand. There is even more of this than the total amount of conventional oil. In order to produce this oil, large amounts of energy from

natural gas (also approaching its peak) are needed and the process is heavy and investment-intensive, compared to the production of conventional oil. The largest reserves of unconventional oil, globally, are located in the United States, and in the state of Alberta in Canada. The Canadian government estimates that by 2025, Canadian production of oil from oil sand will amount to 3 million barrels per day, which is only 4 percent of the current global oil production.[9]

– For natural gas the situation is similar to that of oil. In his book *High Noon for Natural Gas*, Julian Darley, based on extensive research, concludes that the US production has already peaked in 2003 and the global peak will be reached during the coming decade.[10]

The demand for oil has, until now, been growing by a steady 1.5 percent per year, largely because of the economic growth in China, India and South-East Asia. Due to the more recent financial turmoil, the increase in demand has, temporarily, declined. With limited opportunities to increase production and the possibility that production volumes will start to fall during the next few years, we have to expect price increases in the near future, which will influence economic growth. According to the US "Hirsch report," a 50 percent increase in oil prices causes economic growth to decline by 0.5 percent.[11] It is, however, unclear in the report weather this relationship holds at all price levels. What was true when a 50 percent increase amounted to 10 dollars may not hold when the price of oil is 100 dollars and a 50 percent increase amounts to 50 dollars.

The oil industry has failed to address these issues. The industry maintains that Saudi Arabia will increase production, and that deep sea wells, unconventional oil and arctic oil will make up for the decrease from countries that have reached their production peaks. According to peak oil analysts, this claim cannot be substantiated by the available data. Instead, Swedish expert Professor Kjell Aleklett, Chairman of The Association for the Study of Peak Oil (ASPO), warns of oil shortages during the next decade.

According to the Hirsch report, if we start strong mitigation efforts when the peak occurs, we have to expect a shortfall of oil for 20 years. If we start mitigation efforts 10 years in advance of the peak, we have to expect moderate supply shortfalls. It is only if we start mitigation efforts 20 years ahead of the peak that we can expect to entirely avoid supply shortfalls. This is the reason why we, in this book, will not continue to discuss whether peak oil has occurred, will soon occur or will not occur until 15 years into the future. If it has occurred or will soon occur, we are already late; if it will occur 10 years from now, we need to start a crash program; and if it will not occur until 20 years into the future, we are lucky, but it is nevertheless just about time to begin.

It is not for me to judge which of these alternatives is closest to the truth. This must be part of the strategy development effort, which is advocated later in the text, and the level of transformation efforts must be adapted to the urgency of the situation.

With an oil price above 60 dollars per barrel in the winter of 2008 and with no structured argument against the warnings of the peak oil analysts, we have reason to believe that peak oil may be around the corner.

CHAPTER 5

Three Examples of Large-Scale Transformation Efforts from American Twentieth-Century History

The recurring theme of this book regards the opportunity for economic renewal and business development that is represented by the energy transformation challenge that lies ahead of us. In order to drive home the argument that a large-scale effort to transform energy systems is both realistic and could benefit society, we will take a closer look at three, by any standards, amazing successes of large-scale American projects. We can learn many things both from the effort by the United States in the Second World War to become the "arsenal of democracy" and from the space program, which, among other things, put a man on the Moon in 1969. The Marshall Plan, which was developed by the US Foreign Secretary, George Marshall, after the Second World War, in order to assist the free nations of Europe, in their efforts to recover from the war, will also be discussed.

These three government projects could serve as examples both of planned effort in American and global economic and business development and of the large-scale development of new technologies, "products" and services. There are also important differences between these different programs, which we need to recognize, and that we can learn from.

5.1 THE UNITED STATES IN THE SECOND WORLD WAR

Some authors have compared the effort needed to transform energy systems to the US war effort during the Second World War or to the race, by the United States, to put a man on the Moon. Even if I hesitate to compare the, hopefully, peaceful effort of transforming global energy systems to a war effort, there are several parallels to war, since a war effort involves the whole population of a country. In the case of the space program, this involves mainly scientists, engineers and people employed by NASA and its suppliers. In the case of transforming society to a sustainable future, everybody, or most people, would have to become involved and most of us would have to change a number of aspects of how we live and work.

In the case of the US war effort during the Second World War, this involved the whole country and all of its people and it was initiated very rapidly. According to John Steele Gordon, in *An Empire of Wealth*, which is a history of the economic development of the United States, Winston Churchill, as the new prime minister of the United Kingdom in May 1940, sent a very frank demand in a letter to Franklin D. Roosevelt. The most immediate needs of the United Kingdom were the loan of 40–50 of the US navy's older destroyers, several hundred of the latest type of aircraft and antiaircraft equipment and ammunition. In his letter Churchill also asked that Britain be allowed to purchase steel from the United States and also purchase other materials. He promised to pay in dollars as far as possible, but made it clear that he felt "reasonably sure that when we can pay no more, you will give us the stuff all the same."[1]

At the time the military resources of the United States were small. The army of 300,000 soldiers was smaller than that of Yugoslavia and recruits had to sometimes be drilled using broomsticks for lack of weapons.

5.1.1 Resources Buildup

According to Gordon, Roosevelt immediately realized the precarious situation of the United States and asked Congress to increase the defense budget by $1.3 billion and he asked for the production of at least 50,000 planes a year. Later the same year Congress approved to draft 16.4 million men in order to find 1.2 million soldiers and 800,000 reserves to train.

In order to win the war, Roosevelt realized that the United States would have to become "the arsenal of democracy." In the first 6 months of 1942, the US government gave out more than $100 billion in military contracts, which was more than the gross domestic product of 1940. During the 3.5 years from 1942 to 1945, US production amounted to "6,500 ships and boats; 296,400 airplanes; 86,330 tanks; 64,546 landing craft; 3.5 million jeeps, trucks and personnel carriers; 53 million deadweight tons of cargo vessels; 12 million rifles, carbines, and machine guns; and 47 million tons of artillery shells, together with millions of tons of uniforms, boots, medical supplies, tents and a thousand other items needed to fight a modern war."[2]

As an example of how American industry turned completely toward the production of military equipment, Gordon mentions that

The Ford Motor Company alone produced "more war material than the entire Italian economy." This means that this company and other US firms almost completely gave up their civilian production in order to cater to the needs of the war. In 1944 the Ford plant in Willow Run produced one new B-24 bomber every 63 minutes. In contrast, between 1943 and 1945, the American automobile industry produced 37 automobiles.

This tremendous effort not only required the complete readjustment of the US industry, but it also meant that large numbers of women had to work in factories and in a number of other production-related jobs and positions, while their husbands were doing service in the armed forces.

5.1.2   Administrative Resources at the National Level

As we all know, the Allied powers won the war. It was largely won by the United States turning its entire market economy into a planned economy, and this was done in almost no time at all. As Gordon points out, in a capitalist economy, run by consumers, central planning is highly inefficient as a means toward satisfying the economy's needs. It works much better when the goal is to satisfy an army, navy and air force by producing for the more predictable needs of war.

In order to coordinate production efforts, a new authority was created, which Roosevelt called the War Production Board (WPB). At its peak it had 25,000 employees in 1942 and it was the largest of the war bureaucracies. As its manager Roosevelt chose one of the most able managers in US business, namely Donald Nelson, former Executive Vice President of Sears Roebuck. As a vice president of this company he was in charge of purchasing production resources for the production of all goods in the Sears Roebuck catalog. He knew American production inside and out. As a business manager he made $70,000 a year. When he joined the WPB he settled for $15,000.

Running the largest planned economy in the world at that time, Nelson had to focus on three tasks. He had to find out from the US military and its allies what they needed to win the war. He had to inventory the materials and the production resources available to meet this need, and he had to find ways to make up for the differences between supply and demand that would inevitably arise.

5.1.3   Technology and Materials Development

Not only did the war production effort provide the necessary equipment for the war, but it also led to the development of a large number of new technologies and materials. Many raw materials were in ample supply in the United States, but there were necessities that could not be produced in the United States in large enough volumes that had previously been supplied from abroad. One such material was rubber.

The development of a large-scale process for the production of synthetic rubber from oil was invented and production was rapidly expanded. In 1945 DuPont and other companies produced 820,000 tons of this material, a figure that becomes all the more impressive as the production in 1939 was virtually zero. This barely sufficed to cover the military needs and there was little left for sale on the civilian market. Those who had an automobile had to patch their flat tires and wait to buy replacements until the war was over.

Apart from new materials, the war effort also brought forward a number of new technologies that would later be used for civilian purposes. One example was the ability to build lighter and stronger frames for aircraft that were needed for bombers and other craft to carry heavy loads long distances. After the war, this technology was used in the Boeing 707 passenger aircraft and other aircraft models, which soon made the US production of long-distance passenger trains and Atlantic cruisers obsolete.

Another technological invention was the computer, which was needed to calculate the trajectories of artillery shells. The first computers were extremely expensive, very large and energy consuming, but as we know, the predecessors of today's PCs gradually became smaller, more energy efficient and easier to use. The computer and other electronic products, which spawned the microprocessor industry, have, to a large extent, fueled the economic boom of the past few decades. True, the increasing use of energy has been one prerequisite of economic growth, but the ability to communicate on a global scale has, as we will see, made it possible to structure the global economy and global production and supply chains to the level that we are experiencing today.

5.1.4 Individual and Social Consequences

Apart from the military sacrifices that are necessary at a time of war, there are substantial civilian sacrifices that are necessary in order to keep the war machine running. For the sake of comparison between the war effort and the transition to energy efficiency, it makes sense to mention some of the civilian sacrifices that had to be made in the United States during the Second World War. Once more we turn to John Steele Gordon, who describes how unemployment was eradicated overnight as a large share of the male population was drafted and women had to take their places in the production and transportation of war materials and in the production of raw materials for this purpose.

In addition to this there were shortages of goods. Tires were the first product to be rationed and many other products made of rubber were not available in stores during the war. This was also true, as mentioned above, of automobiles and other industrial goods such as refrigerators. In 1945, 13 different rationing programs were running and a number of different products, ranging from gasoline, sugar and coffee, to butter and oils, meat and shoes, were rationed within these programs.

Through the war effort, many careers changed direction. People who had trained for a particular line of business had to spend the war years learning something else and many times people stayed in their new businesses.

One of the advantages for the veterans who survived the war was that they were offered extensive opportunities to get a high-level education. This was done, primarily, in order to keep a number of veterans happy and away from the labor market for a few years, so as not to create a labor surplus. For American society this led to a tremendous rise in competence. More than 8 million veterans received more education in college and technical schools than they would have received otherwise. The percentage of the population that had college degrees increased and the number of college degrees awarded doubled in 1950 compared to 1940, before the country joined the war. This huge investment in training and education, not only of the veterans, but of building and expanding a national education system, has benefited the US economy ever since.

5.2   THE MARSHALL PLAN

After the war Europe was devastated. Production resources, houses and infrastructures were in pieces. According to Greg Behrman in *The Most Noble Adventure,* in many the countries that had participated in the war, people had to live on an average of 1500 calories per day, a level that increases the risk of disease. In Germany 40 percent of houses were uninhabitable, and in Berlin 75 percent of houses were in the same state. The same figure for Britain and France was 30 and 20 percent, respectively. In the same way a large share of the production resources and skilled people had been lost in the war. In the years after the war, very harsh winters and extremely dry summers made the problems even worse. In 1947, France, because of drought, had the smallest harvest in 132 years.[3]

At the same time, the trade relationships and other contacts that had existed before the war had broken down. Germany and Italy had been the enemies of the rest of Europe, and all relationships with these and other countries remained severed for several years.

In this situation, the commander in chief of the US forces during the war, to whom President Truman had given important political assignments after the war, General George Marshall, together with the state department, devised a plan that would assist Europe in its reconstruction. The war had cost the United States 300 billion dollars and 300,000 lives. Now, Marshall and his colleagues suggested that the country should spend another 17 billion dollars, over 4 years, on foreign aid to its former allies and enemies. The plan was to make the countries of Europe responsible for the results of the program, and that the United States should finance a substantial part of it. Another important goal was to recreate a spirit of cooperation in Europe, which was deemed necessary in order to create the foundation for a lasting peace.[4]

Finally, the cost of the plan turned out to be 13 billion dollars, which corresponds, in today's currency, to 100 billion dollars, and in terms of the share of the gross domestic product to 500 billion dollars.[5]

5.2.1 Making a Strategy and Organizing the Program

After Marshall had presented the idea of the plan, at a celebration ceremony at Harvard University, in early June of 1947, it took almost a year until the plan could be put into practice. This year included substantial efforts at analyzing the economic and political situation in Europe, and the needs of aid in the different countries were documented. This was largely the responsibility of the countries themselves, but they were aided in this work by the United States that wanted to be involved in many aspects of the strategy development of the program.

Only a month after the idea of the plan had first been presented, on 4 July, the British Foreign Secretary, Bevin, and the French Foreign Minister, Bidault, in a joint communiqué, invited 22 European countries to participate in a plan to reconstruct Europe. On the next day, the Soviet Union pulled out of the plan, and it advised its satellite states to do the same.

During the coming months of analyses, discussions and political bargaining, in talks between the United States and Great Britain, based on the strategy analyses, it was agreed that one of the success factors of the plan would be the restoration of German production capacity. This was strongly opposed by France, which feared that its old enemy would regain its economic and production strength and resume its series of wars with France. The conflict between Germany and France, and the resulting differences in opinion between the United States and Britain on one side, and France on the other, was the most difficult issue to resolve, apart, perhaps, from the financing of the program on the side of the United States, which met with strong opposition from many, especially Republicans, who thought that the United States had enough of its own internal problems.

In the United States, the program was administrated by the Economic Cooperation Administration (ECA), headed by Paul Hoffman, a former high-level manager at Studebaker, who was appointed in April 1948. When he arrived in office, he found 10,000 job applications and 500 personal letters from all kinds of people, from members of Congress, to Cabinet members, ambassadors, journalists and job seekers, who wanted to see him. For the organization of the agency, and the architecture for how the aid was going to be spent, Hoffman appointed Richard Bisell, a professor at MIT. When he left Cambridge for Washington, Bisell had packed for a 5-day stay, little knowing that he would never return to live at Cambridge.

Bisell was assigned with the task of spending the dollars of the American taxpayers well. There was to be no corruption or scandal in the administration of the program. In order to achieve this, the management implemented a novel procedure, that of counterpart funding. Another of the tasks of the ECA was to make the activities of the program market based, so that competitive bidding for contracts became the main mechanism of the program.

In order to make it possible to administrate the counterpart funds, the contributions of each of the 16 participating countries were deposited in the government's central bank in each country. These funds were earmarked for the purpose of recovery, and the ECA, together with representatives from each participating country, had to develop a strategy for spending these funds. In some countries, they were used in order to pay back debt, and in other countries they were used for industrial investments. The administrative procedures for the financing of actual transactions, such as the financing of an American tractor for a French farmer through the French national bank, and the ECA, were relatively complicated, but they worked.[6]

5.2.2 Setting Goals and Making Plans

After negotiations and planning, in August 1947, at a meeting of a five-nation Executive Committee at the US Embassy in Paris, the Head of the State Department, William Clayton, laid out the terms in the form of seven overarching goals for the program:

1. To achieve self-sustainability for Europe in 4 years.
2. Each year, aid was to be reduced.
3. Europe was to focus on production, primarily of food and coal.
4. Long-term goals were not to interfere with the short-term reconstruction of available production resources that were to a large extent located in Germany.

5. The economic and financial stability of Europe was an important goal.
6. Europe had to liberalize trade.
7. A permanent multilateral organization in Europe that could promote cooperation for the long term had to be formed.

The plan was presented by Clayton as "friendly aid," and he emphasized that the initiative remained with Europe.[7]

In order to ensure that the goal of self-sustainability was met, production plans became key to the program. The most important goal for Hoffman, and the ECA, was to increase Europe's economic output by one-third by the end of the program. This is an example of how the overall goals set down by Clayton were translated into detailed and quantitative goals by the administrators of the program. At the time many people doubted that this could be accomplished. In order to achieve this goal, the ECA, together with the participating countries, developed production plans for each country and analyzed which measures needed to be taken in order to achieve the goal.[8]

5.2.3   The Program

In December 1947, still according to Behrman, the first shipments of food arrived at Le Havre, to be distributed across France. The dock workers of France, who had been on strike before the shipment against American imperialism and the Marshall Plan, returned to work and now offered to work for free with unloading the 1500 tons of food, which filled 10 trains that headed for different regional centers in France. At the same time ships with similar loads had left harbors on the American East Coast for the cities and towns of Italy.[9]

The first aid was officially sent under the banner of the European Recovery Program (ERP) was sent on 14 April in 1948. On this day a number of ships started loading in Texas with a total of 54,000 tons of grain, fuel, foodstuffs, feed, chemicals, fertilizers, raw materials, semifinished products, vehicles and equipment. These shipments were followed by a large number of similar shipments over the ensuing years.

Within the first year, Hoffman had employed 620 people in positions in the United States and Europe. The ECA had authorized 750 million dollars in aid. Seventy-five percent of this amount went to the three largest countries, the United Kingdom, France and Italy. One-third of the aid came in the form of food and feed; the rest was goods that were necessary in order to rebuild morale and production capacity. Many times the shipments had immediate effects on the production capacity of the European countries.

The program, for several reasons, ended 6 months before its original expiration date. During the program, the United States had spent 13 billion of the 17 billion dollars that had been approved by Congress. The result of the plan was better than the goal set by Hoffman and the ECA. "From 1947 to 1951, the Marshall Plan countries' aggregate gross national products grew from $120 billion to almost $160 billion. By the end of the Plan, Western European industrial production was 35 percent higher than the prewar level, exceeding the lofty goal Hoffman had set at the Plan's start".[10]

And, as we know, the plan also achieved the other goals set by Clayton. The economic growth that was largely created through the plan has continued, economic and financial stability has been achieved, and free trade has been one of the tools behind the growth. Another important contribution has been the cooperation in Europe, which has resulted in the European Union, and the fact that Europe is now becoming an economically, as well as politically, integrated region.

5.3   THE APOLLO PROGRAM

The space program is an example of a large-scale effort toward a national goal. In the face of Soviet advances in space, President Eisenhower in 1957, reluctantly, created NASA, and a small group of 45 persons, called The Space Task Group, was formed in November of 1958. In the same year, the President also created the Advanced Research Projects Agency (ARPA) and gave it the task of coordinating advanced projects and formulating national goals. In 1961 President John F. Kennedy challenged his country with the task of putting a man on the Moon, and bringing him safely back, before the end of the decade. He also increased the size of budgets and other resources from the small and cautious start made by Eisenhower. It took only 10 years and 8 months from the formation of The Space Task Group, to accomplish Kennedy's challenge. This was a tremendous endeavor, and it has been estimated that 400,000 Americans participated in the task.

Once President Kennedy had broadcast his challenge, The Space Task Group rapidly finished its Statement of Work that included the specification of the spacecraft, a work that had already been started. Kennedy made his speech on 25 May, and from 18 to 20 July NASA held a conference for the Apollo program, during which it informed potential bidders of the contents of the Statement of Work. Later in July a tender was sent to 12 companies, and these included some of the technical specifications for the Saturn V rocket. Later, the specifications for the units, which were necessary for the landing mission, were to be made. In February and March 1962, a number of contracts were given to major contractors for the key parts of the craft. In July 1962, companies were invited to bid on the development contract for the lunar landing craft, but it was not until the end of 1962 that NASA decided how it would solve all the issues related to the challenge from the President.[11]

The Apollo program included a tremendous number of technical challenges that had to be mastered in order for the mission to succeed. In addition to these, there was also a tight schedule of test flights, which made it necessary for as many as four Saturn V rockets to be built at the same time in the huge Vehicle Assembly Building (VAB). Considering the enormous size of the rocket, which was 380-feet tall, besides its almost equally large "umbilical tower," it would have been logical to assemble both structures in place at the firing pad. Due to the need to be able to assemble four rockets in parallel, in order to allow for four test flights per year during the program, the assembly had to take place at a separate location. In addition to the tremendous technical advances that were necessary in order to develop the solutions and the technologies for the booster, the capsule, the lunar lander, the heat shields, and the control systems, to mention only a few of the parts, the Apollo program also had to include the construction of the VAB, which is the largest man-made structure on earth. This building is 525-feet high, and its foundation covers 8 acres.

In order to transport Saturn V and its umbilical tower to the launchpad, a special movable platform measuring 150 × 160 feet, with one large truck with belted links at each corner, was constructed. Each link was 8-feet long and weighed a ton, and each of the eight belts contained 57 links. The rocket had to be transported in its vertical position from the VAB to the launchpad. In order to maintain its level position all the way, also along the 5° climb to the pad, hydraulics that gradually adjusted the platform as it climbed were used.

In retrospect the achievement of putting a man on the Moon in only 10 years and 8 months from the day Eisenhower created The Space Task Group is amazing, considering its humble beginnings, as a task force of only 45 people. The expansion of the project to encompass a total of 400,000 participants, most of them with clear responsibilities and deadlines for their deliveries, is in itself a tremendous management accomplishment.

While the example from the Second World War represents a total makeover of the American economy, virtually overnight, into a planned economy with one single goal, and the Marshall Plan represents a focused effort of development aid to Europe, the Apollo program, and other parts of the space program represent a major national effort, which has been run in parallel with normal economic activities, and which worked as a powerful force in developing new technologies and the economy as a whole, while at the same time maintaining and developing a strong civilian sector in the economy. In addition to their market-based activities, many US companies worked (and still work) extensively on government orders to deliver technologies, products and services for the space program.

In all three cases, there are strong elements of centralized planning, but all the time, the key philosophical tenets of market economics formed the bases for the developments.

5.4 AN INTERPRETATION OF THE EXAMPLES

Previous authors on peak oil and the challenge of energy transformation have compared the transformation of energy systems to the US World War II effort and the Apollo program. In doing this they had mainly the level of investment and the amount of work in mind. After presenting the three case studies I now wish to take the comparison of previous large-scale projects a few steps further.

5.4.1 Economic Growth and Business Transformation

Three examples of large-scale government projects have been described, through which the United States, during the past 70 years, has contributed substantially toward the fulfillment of a number of major high-level goals. These three examples represent large-scale investments and planned programs, which have helped the market economies of the world to develop. These programs have contributed to the development of the global market economies and they have also contributed to the position of the United States as the uncontested leader of the global economy.

A large number of government-funded programs in different areas have contributed to the development of new technologies, industrial resources and products, which have become important exports for the United States. These development initiatives have also contributed to the development of competence and skills across the United States and in other countries. These skills have become important factors behind the productivity growth that has fueled global economic development. The benefits of the programs have in many ways spilled over to other countries, through technology development, trade and other types of exchange with the United States.

5.4.2 The Aspect of Planning

The three projects described above represent three levels, or degrees, of planning:

1. The Second World War effort is an example of a single-minded focus on one overarching goal – that of transforming US industry

to become the "arsenal of democracy" and leading the Allied forces to win the war. In this case, the planned economy took over "all" economic activities, and winning the war became the only priority in only a short space of time.

2. The Apollo program represents a large-scale national example of a huge development project with the aim of "putting a man on the moon by the end of the decade." This program was run in parallel with other normal peacetime economic activity, and the overall priority of "business as usual" and economic growth were maintained throughout the program.

3. The Marshall Plan represents a program of selective investments in the reconstruction of particular production resources. The US government did not take responsibility for the final result of the economic recovery of Europe. Instead, the Marshall Plan fueled the economic development and unification of Europe, which would probably have taken place in any case, but at a slower pace.

The different examples also show that a number of different mechanisms have been used in the different programs. The United States has employed tools, such as direct investment in technology development, and financing of projects, government procurement of large volumes of products, subsidies for technology development, loans and other tools that have made capital available in areas where it was needed.

We can learn many things from these examples, when it comes to the planning of the energy transition. First and foremost we can learn that the principles of market economy and those of long-range government planning of aspects of economic and industrial development do not exclude one another. On the contrary, they complement each other.

5.4.3   Organizational Learning

Organizational learning is the science that studies how complex organizations cope with new complex tasks and how they adapt to changes (expected and unexpected) in their environments. The three projects discussed above represent examples of highly complex challenges, for which society at that time was not prepared. In this type of situation, in which a certain task is given a very high priority, society is able to adapt very rapidly. This adaptation and the creation of appropriate organizational structures for organized problem solving is the essence of organizational learning. Within the appropriate organizational structures, individuals can learn and be supported by the learning of individuals can organizational tools that capture the structural capital of the organization. Such tools may consist of job descriptions, business processes, IT systems, and tools for analysis and decision making.

While this process is ongoing and beneficial to society, organizational learning in society increases the level of specialization in society and the task of reorganization for a new purpose becomes increasingly difficult. For this reason, we can learn from the above examples, but our present society consists of more specialized organizations and individuals than the societies of the 1940s and the 1960s. During the energy transformation program this will become obvious in a number of ways and in different situations and this will need to be tackled. In order to change today, we will probably need to spend more on change management, in the form of people who coordinate different aspects of the change effort, than has been the case in previous programs.

The disciplines of organizational learning and change management, which are of a relatively recent date, provide the tools that we need to do this.

5.4.4   Government Programs and General Purpose Technologies

In his book *Is War Necessary for Economic Growth?*, Professor Vernon W. Ruttan of University of Minnesota analyzes the question, posed in the title, whether war is necessary for the US economy and other economies to grow. Ruttan himself provides the answer to this question, saying that war and preparations for war may not be necessary, but that large-scale government-funded projects are necessary in order to drive economic growth through the development of general purpose technologies. The six general purpose technologies that are analyzed by Ruttan are as follows

1. interchangeable parts and mass production,
2. military and commercial aircraft,
3. nuclear power,
4. the computer industry,
5. the Internet, and
6. the space industries.

In conclusions to the book, Ruttan argues that the private sector can never be expected to develop the general purpose technologies that are necessary in order to drive future economic growth. The reason for this is that companies do not have access to the "patient" capital, which is needed in order to run 10- or 20-year technology development projects. The only possible source of such capital is the government. Ruttan even believes that efforts by the private sector to develop new general purpose technologies in the biotechnology and molecular biology sectors need support from large-scale public sector projects, in order to produce new general purpose technologies by mid-twenty-first century. Ruttan also finds it less likely, with the lack of a major military threat to the safety of the United States, that future investments of the scale of previous development efforts in military technology can be justified.

Thus, the most probable answer to the question whether war is necessary for economic growth is "No," but large-scale government spending is. In the next few decades, there will be a need for a number of new general purpose, energy-efficient technologies in diverse sectors of society, such as transportation, energy production and distribution and housing. In the face of massive investment needs over the next decades, society will need to allocate very large volumes of capital and other resources to the transformation effort. The total amount of investments will not only have to cover the development, production and implementation of new technologies but also the development of new business processes and the implementation of these in companies. In addition, there may be a need for other investments in public transportation and the remodeling of cities, based on the implementation of new transportation systems.

Since some alarmist analysts of the peak oil issue expect resource wars, because of a shortfall of oil, I need to point out that a large-scale transition to renewable fuels and sustainable energy technologies would constitute a better long-term solution to the problem of an energy shortfall, than warfare to "secure" resources that will last for only a short period in any case. Large-scale transformation efforts would represent both more humane and financially more lasting investment alternatives, than a war effort.

We must hope that the transition could be run in the same way as the space program, but if we, as a society, wait too long with the transformation efforts, we may have too little time available for it to be possible to run the transition program in parallel with other activities. In this case, the program may need to more resemble the single-minded effort of the Second World War than the effort of the Apollo program.

I have chosen the three above examples, in order to illustrate the fact that planned government programs could be given different forms, and that different programs could include different amounts of planning, and different types of goals and mechanisms. Depending on the urgency and the financial constraints on the transformation challenge, different types of programs could also provide different amounts of space for the mechanisms of the market economy to play their roles. Because of the urgency of the situation, the Second World War represents the case in which planning takes the most prominent place and the market mechanisms were given an insignificant role. In the case of the Marshall Plan, the market mechanisms were given a central role. Through the strategy process for a transformation program, we can determine how the program needs to be run in the present case.

5.4.5   Organizing and Planning the Energy Transition

In the case of the Second World War we can see that it was really a tremendous effort in terms of the number of vessels and vehicles built and in terms of the technologies that had to be developed in order to win the war. We can appreciate the similarity of the effort at hand if we ponder the number of cars, trucks and airplanes that need to be replaced with more fuel-efficient ones if we want to maintain our affluent life styles and not simply stop consuming. We can also consider the number of power plants for renewable energy production, production plants for renewable fuels, such as ethanol or synthetic diesel, or biodiesel, and the need to rebuild electricity grids and other installations. In industrial companies existing pneumatic systems and other power sources, in addition to pumps, heating and lighting, may need to be replaced or complemented by more modern and energy-efficient installations. We may also need to rethink distribution networks and the location of production, so that we will need less transportation to bring products and services to market.

We may also, already at this point, understand that we, as a society, may need to make sacrifices in order to turn production toward energy-efficient materials and systems with the speed and determination that is necessary. This may mean that, during parts of the transition, we will have to reduce consumption of different types of goods and services, and that this reduction may in some cases be permanent. It may also be the case that we, to some extent, as individuals will have to work with matters and perform tasks in the transformation program that we previously did not expect, since it is almost certain that many, or most, of us will have to participate in the transition.

Apart from the sheer volume of work and investment that need to be performed, we can learn more things from the war effort. One of these regards the level of organization of the military and production systems. As we have seen, in the Second World War, the market-based capitalist system of the United States was replaced by a system based on central planning, which was necessary and appropriate in that situation. In any case a formal government organization, similar to NASA, the War Production Board, and the Economic Cooperation Agency, will probably be needed.

5.4.6   The Role of Goals

Many management theories testify to the assertion that goals are important for the successful management of economic and business endeavor. According to business theory, high-level stretch goals constitute important focus points for people at all levels of a business development or transformation program. According to business handbooks and methods, such as balanced scorecard methodologies, top-level goals need to be broken down by department and individual, in order to function as operational guidelines for action on a daily basis.

Goals need to be quantitative. This could mean that goals for the reduction of CO_2 emissions and energy consumption could be formulated on national and state levels. These goals could be broken down and interpreted at corporate levels in order to be formulated as quantitative reduction goals for an organization. These top-level goals could then be broken down by department and translated into individual action. Another possibility is to break goals for emission reduction down to the need for the penetration of particular technologies and fuels, and the expansion of production resources and marketing efforts could be planned based on these goals. This is the best alternative for the goals of the companies, streams and projects that will be focused on technology, product and service development.

Sometimes, quantitative goals for the end result cannot be specified. In such cases qualitative goals, or intermediate quantitative goals, may be set, such as the goal of making a certain amount of US government money available for investments in Europe through the Marshall Plan.

Box 5.1 "Big Hairy Audacious Goals"

In the research project leading to the book *Built to Last*, a team led by James Collins and Jerry Porras studied a large number of US companies that had been exceptionally successful for more than 50 years. Among the companies were Boeing, Ford, Philip Morris and a host of other large companies. This team of researchers identified a number of critical success factors that characterized the corporate cultures of these companies. One of them was the fact that these companies were built on the value of setting very tough and demanding goals for the whole organization. These goals were not just goals for sales or profitability but goals for product and business development that really stretched the capacity of the companies and their employees.

The authors called this type of goals "BHAGs," which means "Big Hairy Audacious Goals." Boeing represents the prime example

of a company that repeatedly has used BHAGs in order to motivate and focus the attention of its employees. In 1952, when Boeing
received 80 percent of its income from one customer, the US Air
Force, the company set out to create the first commercial jet. At
the time airlines had the impression that Boeing was a producer of
excellent bombers, and they had expressed little interest in a commercial jet from Boeing. The main competitor, and market leader,
Douglas Aircraft, did not represent a threat in this respect. This
company had no intention, at that time, of developing a commercial
jet. Boeing took on the project in order to create a business advantage over Douglas in the commercial aviation market.

The project budget was three times the annual after-tax profit, and
a quarter of the net worth of the company. Boeing had to gamble
15 million dollars of its own money in order to build a prototype. The
result was the 707 jet, which was the first commercial jet. Douglas did
not present a jet of its own, the DC-8, until 1958, and by then Boeing
had taken control of the commercial market, a position that it still
holds. According to Collins and Porras, Boeing has made the setting
of distinct BHAGs one of its most important competitive advantages,
and it has used it consistently throughout its development.

Another example of a company that has used BHAGs in order
to achieve superior results is represented by GE. Former CEO Jack
Welch set the goal for GE to "become number one or number two
in every market we serve and revolutionize this company to have the
speed and agility of a small enterprise." This distinct goal formed the
foundation of the success of Jack Welch in achieving exactly what
was stated by this goal. The competitor Westinghouse that set the
less concrete goals of "Total Quality, Market Leadership, Technology Driven, Global, Focused Growth and Diversified" has never
managed to follow GE in terms of business growth or profitability.[12]

In the case of both the Second World War effort and the Apollo
program, these programs make up examples of BHAGs on a
national scale. The goal formulated by Roosevelt in 1941 for the
United States to become the "arsenal of democracy" and take a
leading role in the fight against Nazism represents one BHAG. The
goal of putting a man on the Moon before the end of the decade,
formulated by John F. Kennedy, represents another. Later goals of
building a space shuttle that could be used and reused several times
for space travel and the construction of a permanent space station
are two others.

5.4.7 The Role of Financing

The extreme focus of the Second World War transformation and the Apollo projects were possible due to the vast amounts of public financing that were made available for these projects. During the last few years of the Second World War there were no other growing and profitable markets for industrial companies to supply to, than the US Government. Similarly, the Apollo project offered an extremely attractive opportunity for the participating companies to build on their existing competencies and resources and to get paid to develop a set of unique competencies that could be exploited in future growth markets. The financial resources that were made available made participation in these projects an offer that companies with the right resources could not refuse. They were paid to enter a learning curve into areas that they could never have entered based entirely on their own financing.

CHAPTER 6

Gradual Change Has Caused the Need for Energy Transformation

In order to understand how the problems of high energy consumption, and the consequent high levels of carbon dioxide emissions, have arisen, we need to understand how we have developed our habits and changed production systems. According to Richard Heinberg each person in the United States now uses the energy equivalent of 50 slaves.[1] It could be argued, that there are a number of reasons for this, such as changed private habits, longer driving distances to work and the increased importation of foreign goods for consumption and production. It could be argued, however, that the one key reason for this, which explains why we need to travel longer distances to work and why we increasingly import the goods that we need from foreign countries, or supply them through long distance transportation in our own countries, goes back to the general direction of economic and business development during the last century.

This development is based on the idea of competition in markets, and the need for increasing specialization of countries, companies and individuals, in order to succeed in the market economy. When goods and services are becoming more similar to each other, or of increasingly similar quality and performance, price becomes the most important aspect of competition. Fifty years ago, most of what was purchased was produced in the region where it was consumed. There were furniture and clothing manufacturers all over the globe, who supplied products to customers in the regional market. The material for construction projects was largely produced locally or regionally, and the food that was consumed had the same local and regional characteristics.

The drive to improve competitiveness has forced companies to increase the level of specialization. Higher levels of specialization of both white and blue collar workers, and higher levels of specialization of the firms that employ them, have contributed to reducing the number of companies that demand a certain type of specialist. In the same way, other changes, such as the reduced need for secretaries and other skilled employees, have contributed to making it increasingly difficult for many people to find jobs locally. More people need to commute to work, and we also need to travel increasing distances to go shopping in large supermarkets and malls, instead of in small local stores.

In the same way, specialization, and the drive to reduce cost, has reduced the number of possible suppliers of any company. This increasing specialization and all the other aspects of the supply chain transformation efforts described in this section are results of the process of organizational learning that is ongoing in society. Fifty years ago, companies could purchase components and services locally, or regionally, from suppliers with a general set of tools and competencies. Higher level of specialization has made most manufacturers focus on increasingly narrow segments of the market, and customers have also been forced, or tempted, to look for better quality, more features and lower prices from specialized suppliers, who are often located at a distance. The cost of transportation has been so low that this has not been an obstacle to this process. This quest for lower prices goes back to the wish of consumers to be able to get more goods and services for each dollar of their income. As imported goods improve our standard of living, our consumption habits that are linked to global production systems for goods and services consume increasing amounts of energy.

In order to understand in more detail how this situation has arisen, and the mechanisms that are involved, we will look at the development of business, from largely local to increasingly global, over the past 50 years. If we develop an understanding of this historical process, we will also understand the complexity of the task of changing production systems back to local again. Sometimes, this may be easy, but only in rare cases. In most cases, it will take time and require large investments.

An important consequence of the development toward cost efficiency, described below, is the ability of society to produce goods and services for an increasingly large population, using fewer man-hours and other resources, but higher levels of energy for transportation and production. While the energy needed to produce each unit of product has decreased, the total volume of consumption has increased, and with this increased consumption, the energy needed for production of increasing volumes of goods and services has increased accordingly. The resulting productivity improvement has made it possible to fuel economic growth and also spread this to Asia, Latin America and other distant parts of the world. The inhabitants of these parts of the world have now become the suppliers of many goods and services to the Western world, through the process that is described below.

6.1   THE FIRST STEP: STRATEGIC MARKET DOMINANCE

I will now go through the development of business strategy, market dominance and operational efficiency in some detail. This is to explain why we have arrived at the present levels of global energy consumption and also to indicate some of the limitations of large-scale energy reduction through the rapid transformation of industrial supply chains. Readers who are already familiar with this development may skip this part or read only parts of it.

Sixty years ago few managers had a plan for the development of their businesses. There were few universally accepted management principles that managers and companies tried to adhere to. The best business advice may have been to buy cheap and sell dear. The subject of business at universities was in its infancy, and, to a large extent, it was the legal and accounting aspects of management that were taught. Large companies and consulting firms had started to internationalize and to develop their own principles and models, based on experience, but there was little research into strategy and what there was tended to be quite rudimentary compared to today's standards. Business was mainly local or regional, and the competition was not as tough as it is today.

Gradually, through experience, a number of principles were developed, based on how to achieve profitability in a competitive and increasingly global environment. In general, according to theory, a specialized company, with the largest market share in each industry, would be the most profitable. This was, at an early point in time, attributed to the effects of organizational learning that were described by the "experience curve" or "learning curve," which was quantified through studies of US aircraft production in 1936 and emphasized in the 1960s by Bruce Henderson, founder of the consulting company Boston Consulting Group, as a tool for understanding business strategy. This curve illustrates how a company, over time, manages to reduce its cost of producing one unit of a product or service, based on the accumulated experience that is developed within the organization and its environment. The rule of thumb is that each time the "accumulated" volume produced by the same company doubles, the unit cost decreases by some 15–20 percent. Thus, the second unit produced will be 15–20 percent less expensive to produce than the first, and the 2000th unit will be 15–20 percent less expensive than the 1000th, and so on. This principle has been shown to hold to a surprising degree.

The lesson from the learning curve is that the largest company, with a long history in a market, will achieve the lowest unit cost, due to its volume and length of experience. The low cost, however, is not only due to short-term advantages of scale, or the accumulation of knowledge on how to negotiate better prices with suppliers, based on volume. On the road toward leadership, the company would have developed strong relationships with suppliers, customers, technology developers, consultants and other partners. Due to the larger volumes, the leader in an industry would have had resources to develop stronger and more active, in terms of the number and size of joint development activities, relationships than any of its smaller competitors. This will lead to a cost advantage not only inside its own organization but also in the whole network of partners that a leading company has developed. The development activities initiated by a market leader, and its business partners, will range from technology development, machinery development, product development, systems development to process and materials development. More recently, large development efforts between companies and their suppliers have gone into supply chain development, collaboration through information technology (IT) and modularization of products. In these cases, the market leaders have also possessed the resources to be more active and run more projects, than smaller competitors.

In general, when it comes to profitability, only the market leader is usually sufficiently profitable over a business cycle. The second, third and fourth competitors in a market with a number of similar producers will struggle to achieve a sufficient profit to keep its shareholders happy. This, to a large extent, accounts for the fact that market leaders, and their suppliers, have more resources to spend on different types of development projects.

The other alternative for a competitor is to develop a niche position, where a smaller company can avoid direct competition with the leader. This can be achieved by offering a subset of the market a unique product or service with features that cannot be copied by the larger companies, because the niche market is too small for them to bother about. Becoming a niche player, the leading company, and sometimes several small companies, can thrive, even if they are much smaller than the leaders in the main market.

We will look a little closer at the process from innovation to market dominance, in order to illustrate how this process started and how markets have arrived at the present situation.

6.2   STEPS IN THE DEVELOPMENT

6.2.1   Innovation

Innovation is the mother of business growth. We have all been impressed by great innovators, such as Thomas Alva Edison, Gottlieb Daimler, James Watt and Alexander Graham Bell. The companies founded on some of their great inventions and the inventions of other visionary engineers and self-trained inventors, to a large extent, fueled economic growth many decades into the twentieth century.

Among the inventions were electricity, telephony and the combustion engine, which still form the backbone of our society's energy, communication and transportation systems. These, and other general purpose technologies, spread very rapidly over the globe and they made many daily tasks much simpler than they had been previously. People had to spend less time lighting and heating their homes and offices, storing and cooking food and transporting themselves to and from work and even to remote destinations. Even if it took several decades for these technologies to spread and to become inexpensive enough to be acquired by people with modest means, most people in the West now enjoy the benefits of a comfortable life. This has been achieved not only through technical innovations but also through service innovations, which have even further reduced the time spent on cooking and purchasing the goods we need, and for transportation and other needs.

Other more recent innovations that have been hailed as completely new to the world, such as mobile telephony and the computer, are, when examined more closely, primarily used to accomplish tasks that people have been performing for centuries or millennia. The computer is used for communication, writing and calculation, bookkeeping and entertainment, to mention but a few important areas of computing. All of these and most other uses of computers represent merely improvements in terms of cost and time, of how these tasks were performed Before computers (BC). The reduction in cost and time may have dramatically increased the market demand for entertainment and communication, driving e-mails and chat sessions into billions of messages sent every day, but the opportunity to reduce further the cost and time of communication must be limited and incremental, rather than dramatic and disruptive.

Similarly, when accounting is done automatically in business computer systems, accounts can be broken down into smaller and smaller parts, reflecting the cost of doing business area by area, which provides management with much improved tools for managing companies. Also in this case, we have to assume that few further improvements in accounting processes can drive cost much further down. Electronic entertainment is of a later date and has been going through a revolution in the past decades – a change that is still going on. Entertainment is now available to virtually everybody every day, which is in stark contrast to the situation a hundred years ago, when a few annual festivities were the only structured opportunities for relaxation from the toil of everyday life. The cost of viewing a play, purchasing an entertaining game or listening to music is much lower today. The lower cost has also led to a dramatically expanding variety of alternatives and an increasing demand for these products.

As we all know, innovation is still going on, but it is increasingly focused on the adaptation of existing products to the needs of smaller market segments, rather than breakthrough innovations for large consumer markets. We develop an increasing number of inventions and new businesses, but for increasingly small average markets.

6.2.2   Market Penetration

Over the past century, important innovations in consumer goods markets, machinery, materials and transportation have been developed and have spread across the globe. At first, they spread throughout the Western world and Japan, and more recently newly industrialized countries have become eager buyers of these products. New markets, such as China and India, are, from a material point of view, developing at a faster rate than the Western markets did when they were at the same point in their development, as these countries have immediate access to the latest and most efficient technologies. Often they adopt mobile telephony immediately, instead of taking the route over telephony across fixed networks, and they adopt personal computer (PC) technology, without first having to adopt mainframe computers.

The markets for products such as cell phones, PCs and cars are growing very rapidly in these new countries. As demand in these countries are satisfied, it will become increasingly difficult to find new geographical large-scale opportunities for market expansion.

6.2.3   Acquisitions and Market Dominance

The strife for market dominance is one of the strongest forces behind organic growth and mergers and acquisitions. However, we can now see that the opportunities to grow further, through acquisitions within a certain market, are limited in many cases. As the global market in many industries is already dominated by a small number of global companies or consortia of partners, there will be less opportunity today than 10 or 20 years ago to drive concentration further by continuing to acquire competitors. Twenty or thirty years ago companies were to a large extent national or regional. The process of acquisitions, since then, has led to a situation in which global markets are dominated by a small number of global players.

Previously, in each market there were a number of smaller companies that could be acquired by the more successful and profitable incumbents. Today, acquisitions have formed relatively concentrated industry structures and tried and tested management principles are applied in most large companies, making it in many industries increasingly difficult to identify acquisition targets where substantial improvement opportunities remain to be realized.

We now have a situation in which many global markets have arrived at a high degree of concentration and the global leaders are represented in all or most of the major markets of the world. This means that some opportunity for growth remains in more recently industrialized countries, such as China, India and Russia, and that there will also continue to be an opportunity in other markets for organic growth at a lower level than previously. However, the large new business opportunities that have been opened up through the structuring of global markets and the opening up of giant markets such as Russia, China and India, that, for different reasons, had previously been closed, will have to be considered to be one-off opportunities. It is unlikely that opportunities of a similar size will appear in the future.

In our society and economic system, we are in a similar way highly dependent on economic growth, which has historically been achieved through productivity improvement. As has been described above, companies have expanded their territories in order to achieve a dominant position in their markets and this has led to economies of scale and other opportunities for growth. We must assume, however, that the average size of future opportunities will be smaller than those of the past. Fortunately, the opportunity to develop new sustainable energy technologies represents an entirely new area for innovation and economic growth.

In the present situation, we are beginning to exhaust the opportunities for further growth in terms of the size of geographic markets. The remaining large and largely undeveloped markets of the past, China, India, Russia and Southeast Asia, to mention the most important, are about to be conquered by global and national companies. There still remains an opportunity for growth in these markets for decades into the future, and, at least in many markets, a potential remains to further increase the strategic dominance through acquisitions and specialisation. This opportunity, however, is not infinite and the returns from such endeavors will diminish. In the future, companies will find fewer opportunities to invest their money profitably in acquisitions that are aimed at market dominance. As we have argued above, the same is true for future investments in new product and service development in maturing industries and business areas. The largest improvements to our daily lives, in terms of time and cost reduction through automation of household chores, daily transportation and tasks at work, have also been made. The average size of cost and time reduction and, accordingly, the markets for products and services that remain to be developed will be smaller.

6.3   MATURE IDEAS

History tells us that the opportunity for society to thrive on a particular idea is limited and that it is often very difficult for a society, which has built its structures, economic system and politics on the pursuit of goals that have been relevant in the past, to move over to growth that is based on a new set of ideas.

In our analysis we can see that the level of specialization and complexity have increased over the past century. First, we had motorcars and people could get any type and color they wanted, as long as they wanted a black Ford of the model T. Now, we have SUVs, sedans, sports cars, compact cars and vans. A few decades ago, popular car models sold in the millions. At present, even the most popular models often sell less than 100,000 units per year. Instead of trucks for the general market, we have light trucks, heavy trucks and trucks for particular purposes, such as for the transportation of liquids, building elements, window panes and other specialized purposes. The global leader in the truck industry, Volvo, may introduce a range of trucks that are fueled by a number of different biological fuels. Many car companies already have cars with engines on offer, which make use of renewable fuels or new engine technologies. This adds yet another dimension to the already vast number of configuration alternatives that are available in the market for trucks and cars. As we increase the specialization and address further small niches of the car and truck market, the number of customers for a particular type of car or truck will, by necessity, on average become smaller than previously. This is a fact of life for companies as of today, but it is also a challenge that increases the cost of development and production of vehicles, which we need to keep in mind when we contemplate and analyze different possible solutions. In order to make the development of new energy-efficient technologies and technologies based on renewable fuels viable, companies will need to, as rapidly as possible, increase production volumes, and reduce production cost for the new alternatives, in order to make the new technologies affordable to customers.

One such idea, which we need to examine closely in the near future, will be our dependence on fossil fuels, and the alternative routes toward sustainable energy systems, the theme of this book. Another idea, which is closely related, is that of economic growth through cost reduction and productivity improvement. As our remaining resources become scarce and the price of resources increases, we need to find new ways of creating economic growth, which are not dependent on the increasing consumption of scarce resources.

When companies had pursued the idea of strategic market dominance for some time, a set of simple principles for the improvement of operational efficiency were developed. During the past two decades these ideas have been pursued vigorously by most large companies, and this development, as well, has contributed to an increasing need for transportation and the use of increasing amounts of energy.

6.4   THE SECOND STEP: OPERATIONAL EFFICIENCY

In the minds of today's managers, the saving of time in a business process is equal to saving money. The saving in terms of dollars is roughly equal as a percentage of the initial cost to the saving in percent on the time it takes from the start of a "production cycle" until it is finished. This principle was first introduced to a wider audience by a consultant from Boston Consulting Group, George Stalk, who had applied this in a number of companies, before he published his ideas in a seminal article in *Harvard Business Review*. The title of the article was "Time Compression: The Next Source of Competitive Advantage" and it was published in 1988.

Initially, this article attracted substantial interest from both academia and practitioners and the book that followed was called *Competing Against Time*, co-written with fellow consultant Thomas Hout. This book became a best seller. Time compression was taken on by companies, initially on a small scale, being used to reduce the amount of capital tied up in operations and the time used for internal transportation of goods within a plant. It was also introduced as a low-tech principle, based on Japanese management principles that were developed by Toyota and other Japanese companies in the 1950s. Thus, Stalk did not invent the principles, but he packaged and labeled them and started to market them to Western companies under the banner of "time compression."

We will now take a look at the history of operations improvement and a view of what remains to be done in the future. The main argument and the reason that we include this detailed review of these tools and the historic economic growth in the book are that we need to understand the process that has formed the current global production and distribution structures, in order to identify appropriate solutions. It is probable that it will take longer to achieve dramatic changes in global supply chains, than to introduce new vehicle technologies and fuels, provided that these are based on cost-effective technologies and principles.

6.5   STEPS IN THE DEVELOPMENT

6.5.1   Operations Research and the Japanese Miracle

The development of ideas in the field of operations research in industry started after the Second World War, building on ideas that had been developed during the war by military analysts. When these methodologies were applied to complex industrial problems, the conclusion was that optimal solutions were usually found through the attuning of a number of mutually contradictory factors. This means that if a company wanted to minimize the volume of inventory in production and the supply chain, this had to be done at the expense of increasing the cost of some other factor. The reduction of inventory usually meant longer delivery times on average, because it was believed that, by necessity, a reduction of stock added up to more stock-outs, an increasing frequency of production halts and less satisfied customers. Similarly, improving the quality of products that came out of production meant a higher cost in quality control, or higher production cost, because workers had to spend more time checking and double-checking that they had gone through all the production steps in the right way, or quality controllers had to be employed in order to test a large number of aspects of the quality of products before delivery, and repair or adjust those that needed adjustment.

For several decades, managers and business analysts alike believed that there were no simple rules of thumb to go by in order to optimize production and other business processes. Meanwhile, in the early 1980s, there was a substantial effort among the same groups of people to decipher the Japanese "wonder." Japanese companies were rapidly winning market share from American and European companies. Initially, this had been true in industries that sold products of relatively low value, such as toys and other less advanced products. During the 1970s, Japanese companies started to threaten both market share and profitability of the incumbents in more advanced industries, such as consumer electronics, machinery and cars. Western managers were baffled at this development, because it had been widely believed that Japanese companies could not develop competitive advantages in high-value sectors of the economy.

At this time, the debate regarding the reasons behind the Japanese miracle was intense. The Japanese culture was discussed: The job security and lifetime employment of Japanese companies differed completely from, especially, the practices in the US labor market. The attention to quality and the general efficiency of Japanese employees were also noted. Some practices of Japanese companies trickled down to the West. The concept of total quality management was developed based on works by Philip Crosby, Joseph Juran and W. Edwards Deming. In his book *Quality is Free*, Crosby argued, based on experiences from Japan, that the quality level of production could be increased, without adding cost to the process. Instead, through standardization of procedures, making routines simpler and training and empowering employees to do each task the same way every time, and solve problems that arise in their work, quality could be increased and cost could be reduced at the same time. Managers no longer had to choose between high quality and low cost. Both could be achieved.

In a similar way, a few years later, in the article mentioned earlier, George Stalk argued that the total time of a production cycle could be reduced and the cost of performing the work would be reduced to the same degree. This was also based on experiences from Japan and contrary to the previous beliefs of managers and analysts. The principles behind "time compression" or "total cycle time compression" had been developed and documented by Japanese companies such as Toyota. To a substantial extent, the reason behind the success of Toyota in reducing cost and working with low levels of inventory was the necessity of keeping capital cost low, due to lack of financial capital. Instead of producing a long series of identical cars, using inflexible processes, as the situation was in the Western automotive industry, Toyota had to produce small series of vehicles and rapidly get paid for them. This gave rise to the principles that made it possible, using low-tech information tools and administration principles, to produce individually customized cars, at a price that was very competitive, compared to American and European alternatives. The management principles that have now been deciphered and described in detail by more recent authors, such as Jeffrey Liker in the book *The Toyota Production System*, contributed both to consistent quality and competitiveness in a number of other respects. Besides the management principles that have been given names, such as lean production, total quality management and supply chain management, Liker adds the aspect of the company culture of Toyota, which adds consistency and focus to everything that Toyota does.

In the following review of some of these principles, particular focus will be given to the aspect of how the system that a company could work on could be extended. We will see how the principle of time compression first was applied, more or less, to sequences of tasks, how it was extended to comprise complete cross-functional business processes and how it later has been applied across complete supply chains. Throughout this development it has been the principle of jointly reducing time and cost that has led the way. Companies that have been doing this for 15–20 years find that the cost of going one step further, for instance in their supply chains, to second- or third-tier suppliers, increases compared to previous steps, and at the same time the average value of the next improvement decreases. This work to improve supply chains, create well-defined interfaces between steps and companies in the chain, and to apply IT business systems in order to facilitate structured communication between partners that work at a distance from one another, has contributed significantly to the globalization of supply chains, and to the increasing need for transportation.

6.5.2   Time Compression

Time compression was introduced by Stalk as a universal principle for the simultaneous reduction of lead time and cost in business cycles, such as production or product development. Stalk used examples showing that time compression could often save 80 percent of the time and cost of a cycle. The saving, it was argued, could be achieved in virtually any industry and it could be achieved for all cycles. A cycle could be similar to what is today known as a process, but it did not necessarily run across the whole business. At this point, companies often worked on smaller segments of the whole process. Lead time was the time it took from the beginning to the end of a cycle.

It turned out that when it was applied across a group of companies, such as the ABB group, where project T50 was run for a number of years, it was sometimes possible to achieve quite spectacular results in a very short time in a cycle from order to delivery. In one unit, producing switchgear, savings of 90 percent on time and cost were achieved and this was used as a model for other companies in the group. In other businesses, however, it was more difficult to save time and less could be achieved. The total savings goal of 50 percent on average on lead times in the group was not achieved during the project. This was due to a number of factors, such as the need to implement new information systems, make changes at suppliers and make other facilitating investments, in order to achieve dramatic savings. The general argument that savings in terms of time and cost were proportional to each other was, however, not disproved.

Stalk used examples of cycles both from order to delivery and from product development to drive home his argument. When a substantial saving was to be achieved, a detailed description of a cycle was made and opportunities to reduce cost were identified. Normally, as Stalk illustrated, substantial savings could be made simply by moving order information, or other information, more rapidly from one department to another. At the outset of a project, in many companies, an order was kept waiting at several stages from the salesperson through production planning, purchasing and delivery of the final product. Just by removing waiting times for the order and moving each single order down the line as rapidly as possible, substantial savings could be achieved.

In the same way an early example of time compression in product development activities was found in the "war" between Yamaha and Honda in the motorcycle business. The war was started by Yamaha in 1981. In this year Yamaha opened a new factory that had the potential of making this company the world's largest manufacturer of motorcycles. Honda, the leader in the global market, decided to retaliate and move resources away from cars and into the defense of the motorcycle business. According to Stalk, Honda and Yamaha both had 60 motorcycle models at the start of the war. In a fierce attempt to outdo Yamaha, Honda turned over its product line twice in only 1.5 years, introducing or replacing 113 models in the process. During the same time, Yamaha, not possessing the time-based work practices of Honda, managed to make only 37 model changes. The speed of development made motorcycle design a matter of fashion and Honda forced the management of Yamaha to surrender, issuing a public statement that the company would in the future recognize the leadership of Honda and not challenge it.[2]

It may be difficult today to understand the true significance of the time compression innovation, compared to previous theories in operations research. Together with the ideas developed in the quality movement, a number of relatively simple prescriptions for efficiency improvement opened up for the spreading of these ideas in the 1990s and early twenty-first century. The interest of business managers in integrated business systems may not have been as strong as it is now, had it not been for the insight that a reduction in time brings with it a reduction in cost that is in proportion to the time reduction.

6.5.3   Business Process Reengineering, Lean Management and Change Management

The now-so-popular idea of business process reengineering (BPR) could be seen as an extension of time compression. The authors of the book *Reengineering the Corporation*, Michael Hammer and James Champy, also added a strong element of IT to the discourse, which from the perspective of spreading the ideas, contributed substantially to this process. Through a number of examples in their book, Hammer and Champy showed that the use of IT for the distribution of information through a business process could facilitate dramatic improvements in performance. Among other examples, they used the example of a process for evaluating loan applications at a bank, which had previously taken several weeks to perform. The reengineered version of this process took only a few hours, and it depended to a large extent on the implementation of IT systems.

The introduction of IT into the equation caught the attention of IT companies. Previously, time compression projects had primarily emphasized the low-tech information exchange tools that had been used by Toyota and other Japanese firms, such as "kanban" signaling through markers on trays for components that were sent between component suppliers and their customers. This signaling was used in order to signal that a tray was empty and had to be refilled. IT solutions for the electronic communication between business systems, called electronic data interchange (EDI), were also used. With BPR, the focus turned toward integrated business systems that distributed information throughout a company and made sure that each decision maker had access to the exact information that would be needed to make the decision. Prior to the implementation of integrated IT systems, systems were developed and implemented for the needs of each particular department in a company. When information had to be transferred from, say, the production department to the purchasing department, this had to be done using a specially programmed link between two systems, or, if the cost of programming and purchasing the software could not be justified, the information had to be manually entered into the purchasing system, using a list that had been extracted from the production system.

This need of manual transfer of data between companies and systems created a need of geographic proximity to other employees and to other companies in the supply chain, in order to ask questions, print lists from one IT system and transfer the information to another, and so on, limiting the opportunity to work closely together with people in remote locations. In this, and many other cases, the interface between the companies, or departments in the same company, was not well defined, and personal communication between employees was needed in order to perform daily routine tasks. This obstacle has largely been removed by integrated IT systems, in which all employees use the same databases to perform different tasks. In this new situation, business units, which work closely together with each other, can be located in different parts of the world and still exchange information in real time.

Integrated systems not only disposed of duplication of work, they also made it possible to speed up processes and even production to cater quickly to customer orders. This practice, which was also introduced in Western automotive companies, was also taken from Japan, where it had contributed both to low cost and to increased revenue. The virtues of time compression were highlighted in a large research program run by professors from MIT, in which some of the principles of lean management were formulated. The results from this project were published in 1990 in the book *The Machine That Changed the World* by James P. Womack, Daniel T. Jones and Daniel Roos. In this book, the practices and performance of European and American automotive companies were compared to those of their Japanese competitors. The study found that not only did the Japanese use less time in all the measured processes, they also organized operations using fewer resources. They found that the Japanese organizations were "leaner" and they urged American and European companies to organize in a similar way.

In a revealing table comparing a General Motors (GM) plant in Framingham to a Toyota plant in Takaoka, the authors showed the substantial differences in time and cost between the plants. They also argued that the principles of mass production used by GM and other Western companies had to give way to the lean principles developed by Japanese manufacturers. It was also argued that the principles of lean production to a large extent were the explanation of the Japanese miracle.

In a comparison between Framingham and Takaoka, the average time used for assembly of comparable cars was 40.7 hours at GM and 18.0 hours at Toyota. After adjustment of these figures, the timings for GM ended up at 31 hours and for Toyota at 16 hours. The number of assembly defects per car was found to be 130 at GM, compared to 45 at Toyota, the assembly space per car was 8.1 square foot per vehicle per year at GM compared to 4.8 at Toyota. In all, GM kept 2 weeks of inventory, compared to 2 hours of inventory at Toyota.[3] The principles of lean production were rapidly translated into a number of principles, which started to be applied at American and European companies. Examples of these principles are "just-in-time" deliveries, "zero inventory" and "zero defects."

It turned out, however, that the transition from mass production to lean management was not as smooth and simple as companies initially tended to believe. It was often a matter of changing the whole layout of plants, retraining employees and management and introducing new measurements and control systems into production, purchasing and marketing. Change efforts involved these changes in the businesses, but they also involved the implementation of new integrated information systems, new machinery and assembly lines and other time-consuming activities. The transformation was not a matter of weeks or months, but a matter of years or even decades. Companies, such as ABB in its T50 project, had discovered the complexity of change. In the case of this project, and, similarly, in other projects, a small number of initial successes could be achieved, which indicated the tremendous potential of such efforts. It was, however, a long way from the beginning of a project to the final achievement of having changed all, or most, processes from mass production principles to lean principles, and achieved the cost reduction and time compression gains accordingly.

These insights gave rise to the competence area of "change management," which combines the different technologies and management principles that are used in order to achieve change, with the understanding of how human beings react to change and how rapidly a change program could be driven in a corporation. Based on the insight that change management is complex, and that a substantial number of people in the organization need to be involved, it has also increased the size of projects.

Typically, a strategy or time compression project in the late 1980s or early 1990s involved a team of four or five strategy consultants who reported to the management team, who took over responsibility for driving certain aspects of the strategic change suggested by the consultants. At the end of the twentieth century a change program in a company such as Chase Manhattan Bank or Telenor, the Norwegian telecom operator could directly involve several hundred persons, who participated in the project on a full-time or half-time basis. Such a project may have involved a strategic reorientation in conjunction with the operational changes that were necessary in that process.

Change management has become a substantial component in large-scale-change projects, both in the public sector and in private companies. Nowadays, a strategic reorientation in one line of business, or the implementation of an integrated business system, coupled with process reengineering efforts, may involve investments of hundreds of millions of dollars over 5 years or a decade. BPR, lean production, change management and other principles for change have substantially increased the market for business and IT consulting.

A corporation such as GE, which over two decades has completely changed its business from mature technologies and products to high-growth and high-tech products and services, would have spent billions of dollars in this process, not counting the investments that have been made directly in the acquisitions or development of new businesses. These investments in acquisitions, process development, IT and change management have accounted for a substantial share of investments in business during the past decades. This structure will never be completely finished, but similar to a railway network, the backbone structure, the control systems and the routines for running business in an efficient way will soon be in place to a large extent. When this happens, we will need new areas to attract our investments that can pay off better than the remaining small operational improvement opportunities in the global business of the future.

6.5.4   Increasing the Scope of the System: Supply Chain Management and E-Business

The time compression movement began with subsets of business processes and continued, under the banner of BPR, with front-to-end business processes in one single company. Some years before BPR was introduced, however, electronic interchange of data between companies started the trend toward what is now called "supply chain management." EDI was first introduced on a large scale in the automotive industry, where companies that wanted to run just-in-time deliveries and other principles influenced by Japanese practices saw the need for low-cost-structured communication with suppliers. Before the Internet and e-business, EDI was comparatively expensive to set up and the cost could be justified only for the automation of data exchange in situations where large volumes of information were exchanged and where substantial savings opportunities could be identified. This was the case in the automotive industry, where it was clearly indicated that the quest for competitive advantage in the future would require a complete reorientation of business practices toward the principles of lean management.

Through EDI the cooperation between automotive companies, and their suppliers, could be intensified and large volumes of data could be exchanged directly between business systems. This exchange related not only to orders, order confirmations, and invoices, but also to production and purchasing forecasts and other information, which had previously been exchanged periodically. An example from one automotive company illustrates how EDI revolutionized both data exchange and work practices. It also shows that supply chain management and BPR are not only about moving materials and finished products faster along an assembly line. The following case example is related to the exchange of production forecasts.

Prior to the implementation of EDI of forecasting data, one automotive company sent one 6-month production forecast each month to its 2000 suppliers. It then took about 2 weeks at each supplier for a secretary to enter all the forecast data into the business system of each supplier in question. The printing of the forecasts and the dispatch of them at the automotive company took 1 week. When the EDI messages for forecasting had been implemented, the automotive company sent one 6-month forecast per day to each of its 200 remaining first-tier suppliers. The cost of transmitting the forecast was virtually zero once the communication link had been set up. In the case of the old process the cost of the data exchange was substantial and the data were outdated once they were entered into the supplier's system. With

EDI, the data at all first-tier suppliers are always kept up to date at a very low cost.

The practice of sharing information with partners in the supply chain is vastly superior to the older practices of secrecy, in order to create uncertainty and short-term advantages in negotiations, between business partners. When actual demand and the best possible forecasts are shared down the supply chain, the effect known as the "bullwhip" effect on stock levels at second-, third- and fourth-tier suppliers can be avoided. The bullwhip effect means that small variations in demand at the company that produces the final product of the chain translate into very large fluctuations in stock levels and demand at the companies further out in the supply chain. This increase in stock throughout the chain increases the total production cost for the final product, which, in turn, reduces the total demand for the final product and for all companies in the chain. Collaboration through the supply chain, using EDI, or the Internet, or some other network for communication, reduces cost and improves profitability. As many authors have observed, it is no longer sufficient to be an excellent company in isolation; in order to succeed you have to be part of an excellent supply chain, and the best way to achieve this is through openness with suppliers and other critical business partners.

Collaboration in the supply chain could involve the sharing of different types of data, from point-of-sales data that is day by day, or purchase by purchase, transmitted down the supply chain, to stock levels at suppliers down the chain, and automatic triggers to supply a certain amount of product to customers down the chain.

The exchange of data nowadays is usually not performed via the old type of EDI communication. With the use of the Internet, the cost of setting up communication links for EDI and the programming of exchange protocols in business systems and communication devices has been reduced dramatically, and it is no longer feasible only for the largest companies to implement electronic exchange of data. Today, all kinds of communication networks, from mobile phone networks to broadband communication, are used in order to facilitate speedy data exchange.

In the minds of present-day managers, e-business not only means the manual placing of orders by individual customers over the Internet. E-business also means the electronic exchange of all kinds of business information between individuals and companies or between two business systems at different companies. E-business is not only used in order to search for the least expensive supplier of a gadget, a book or a collector's item. It is also, perhaps most prominently, used for the very frequent, fast and inexpensive exchange of business information between long-term business partners. In several ways, it facilitates time compression and cost reduction at all companies in a supply chain and it contributes to making the whole chain as efficient as if it was run by only one company throughout.

Despite this high efficiency in terms of time and cost, raw materials, components, and finished goods may circle the Earth several times before the final product is delivered to the customer. This leads us to the core problem that arises from global supply chains – the high consumption of energy in transportation and the accompanying sustainability problems.

If we look at the development of supply chain management in terms of different magnitudes of change, which is of some importance for the coming discussion of sustainability, we find that the development in any company has passed through one or more of four steps: supplier selection, supply chain collaboration, supply chain reorganization and supply chain relocation.

6.5.4.1 Supplier Selection

Supplier selection is a fairly simple form of supply chain management. This means, basically, that a company defines its product or service, often divided into components, and it selects the most appropriate supplier of each component. A number of decades ago, a company typically selected suppliers located in the same country or region, where the customer himself was located. This was for several reasons. First, the practices surrounding global business, such as communication over long distances, and reliable transportation systems, were not as developed as they are today. Second, the tools for communication, such as telephony and IT, were not as developed. Third, the level of language skills was not as high as it is at present. For these and other reasons, it was, for most companies, less expensive 50 years ago to do business with nearby companies, than to seek lower prices from companies further away.

With the development of communication technology, language skills, global business practices and transportation, it has become less expensive to increase the distance to suppliers and select companies that are located in other parts of the world. The gradual development of new supply chain structures has also contributed to learning in this respect. Many times this facilitates low-cost production and other advantages. The low cost of energy, combined with the availability of alternative suppliers of a similar quality in low-cost countries, provides the basis for global sourcing in many industries. In industries that to a lesser extent rely on low-cost production, transportation distances are also substantial, since specialization in these industries has led to the concentration of purchases to a small number of producers in more developed countries.

The Internet has also provided a medium for searching and finding suppliers in remote markets. This could previously be relatively expensive and it was also more expensive to evaluate the quality and reliability of foreign partners. Electronic communication of all types of messages has contributed to lower the cost of the complete supplier selection process, and the low cost, and standardization of messages of electronic communication, with suppliers, at all times, have improved the quality of communication throughout all phases of a business relationship. It is now possible even for small and medium-sized companies to purchase advanced products and services from far away.

Supplier selection may not mean anything more than working together with remote suppliers in the same way that companies previously worked with suppliers next door. In many business situations, especially involving companies whose businesses are not so complex, such as many small and medium-sized companies, this level is sufficient in order to substantially reduce cost. The ability to work efficiently with remote suppliers has, however, contributed to the temptation or need for many companies to reduce cost and, at the same time, increase the consumption of energy in transportation. Despite the fact that regular day-to-day communication between business partners is done in a cost- and energy- efficient electronic manner, there is still a need to, periodically, meet in person. Often a number of people from each company in a partnership need to maintain personal relationships, through visits, which has also contributed to the increasing need for long-distance air transportation. Price reductions on airline tickets have further increased the demand for flights, both for business and leisure.

6.5.4.2 Supply Chain Collaboration

We have already described the essence of supply chain collaboration. Compared to supplier selection, supply chain collaboration means larger investments in frequent communication between business systems. Such investments may be justified in cases where the flow of goods and information is so large that fairly large investments in complex interaction can be paid back within 6 to 12 months.

Supply chain collaboration is applied in many different industries, such as automotive, where, due to EDI, this has been in place for a long time. Other industries where it has been successfully applied are various retailing industries, where companies such as retailers Wal-Mart and Tesco make stock levels of each supplier's products in each store available to the supplier in question, so that each supplier automatically can restock when needed. An increasing number of industries and companies apply EDI for different purposes. In a similar fashion, Toys'R'Us and Benetton have for a long time made point of sales data directly available to suppliers, so that these companies upstream can produce additional amounts of popular items as soon as customer preferences become known.

6.5.4.3 *Supply Chain Reorganization*

Before EDI, each automotive company used to have more than a thousand direct suppliers of components. Each of these suppliers sent their components to the assembly plant for the final assembly of the vehicle. The automobile companies, and their final customers, had to suffer the cost of carrying excessive stock of parts and semifinished and finished goods.

Over the past 20 years, the automotive companies have reorganized their supply chains. They are now supplied by a small number of first-tier suppliers. Second-tier suppliers supply the first-tier suppliers with components, and third-tier suppliers supply the second-tier suppliers with subcomponents or raw materials.

In these chains, the first-tier suppliers supply complete modules for the vehicles. Such modules are, for example, seats, transmission, engine, gear box, brakes and other complete subsystems. These modules are also supplied on a just-in-time basis, allowing for the minimal volume of stock in the assembly plants.

This is a typical example of a complete reorganization of the supply chain. The change has also involved the complete redesign of the vehicles. As we have indicated above, this change has had less to do with the exterior of the cars, which has changed for reasons of fashion and design, than with the reconstruction of the entire car, system by system. This work has taken two decades and it has not yet been completely finished. At the same time, the performance of the supply chains has improved dramatically. The time for assembling a car has been cut by half and the time from order to delivery from a first-tier supplier has gone from weeks down to hours. At present the first-tier supplier starts to produce a module only a few hours before it is assembled onto the vehicle.

A process similar to this has been gone through by companies and supply chains in other industries as well. Many industries with product flows of significant volume have found that the cost advantage of modularization and reorganization of supply chains has paid off very nicely. This is true in industries such as those of computers, where Dell has led the way toward less capital intensive supply chains. Tetra Pak, the world's largest packaging company, has modularized its machines along the same lines as those of the automotive industry, and reduced delivery times from more than 6 months, down to less than 2 months. The same is true for many global incumbents in different industries.

6.5.4.4 *Supply Chain Relocation*

The automotive industry has not only reorganized its supply chains, but the suppliers in these chains have also relocated close to the assembly plants. In extreme cases, such as in the case of the assembly plant for the SMART car, a number of suppliers are actually located in the same plant complex as the assembly plant. In a typical situation 70 percent of the value of a car is delivered from suppliers located less than 60 minutes from the assembly plant. This allows for rapid and reliable deliveries.

The changes described in this chapter, in any one company of some size, represent investments of tens or hundreds of millions of dollars in change management efforts over decades. The transformation has dramatically reduced the availability of local and regional alternative suppliers to the ones that are used today by producers of consumer products, or by their first-tier suppliers. It is, therefore, not so easy, or inexpensive, to change back to previous supply chain structures and to old ways of doing business. The process of change, which has taken decades, may take the same time to reverse, and the financial benefits of reversing the process may not be as obvious as the financial benefits were of running the process in the first place. Therefore, it may be difficult for many companies to financially justify the investments that would be necessary, if we were to recreate local or regional supply chains once more.

Even if in many industries first-tier suppliers are sometimes located close to plants, which may be seen as a localization of supply, second- and third-tier suppliers are still spread all over the globe. In the same way, the final products are increasingly shipped to remote markets from a small number of plants assembling or producing them. For this reason, it is not easy to identify the most relevant changes that would, in the short term and dramatically, reduce the energy consumption and carbon dioxide emissions of these chains.

As the different steps of this supply chain transformation process have taken place, the previously existing local and regional suppliers have, mostly, grown by increasing their level of specialization, remained small or gone out of business. This development has reduced the number of locally available suppliers, and the economies in developed countries have largely turned to qualified services, or highly specialized, automated production. Due to this, it is no longer possible to easily and rapidly reverse the process of supply chain transformation, so that large volumes of goods could be produced locally or regionally.

6.5.5 Specialization and Outsourcing

Twenty years ago, before time compression, BPR and e-business, companies consisted of a large number of different functions. These functions were not very clearly defined inside companies, and the interfaces between companies were also unclear. It often was as if companies had been organized in a slightly random fashion. Each company had also developed certain unique, competitive advantages based on its history and different factors in its environment.

One factor that gave rise to certain skills was demanding customers. Companies that worked with more demanding customers usually ended up with a more competitive set of skills in areas in demand by these customers than the companies that did not enjoy the benefit of demanding clients. It is widely accepted by businesspeople and scholars that the fact that Sweden, with only 9 million people, is the home country of two strong truck companies – Volvo and Scania – is due to the circumstances in the north of Sweden that have given rise to a market for trucks with better performance compared to competitors supplying to customers in less harsh environments. In this case the competition between two strong competitors in the same market has contributed to the success of both companies.

These businesses also grew up together with a number of other more or less related businesses. In the case of Volvo, the truck business, Volvo Car, Volvo Penta, which develops and markets boat engines, and Volvo Construction Equipment started out in the same company. Initially, all functions were also entangled in the same organization. All businesses were served by the same marketing department, purchasing, production and other functions.

In the 1950s and 1960s, following the lead of American giants such as GM and DuPont, large corporations all over the world started to disentangle their various businesses, departments and support functions. One of the first such trends was "divisionalization," which means that large companies with several different businesses started to organize them in different business units, called divisions. Each division was to a large extent a self-contained unit of its own, sometimes sharing some centralized functions with other divisions. The role model for this development was GM, which organized its divisions along the different brands of Chevrolet, Buick, Oldsmobile and Cadillac.

This process has continued to this day, interrupted now and then by the idea of diversification, which runs in the opposite direction, and most Western companies have gradually increased their focus on a small number of core competencies. The computerization within the organizations and between one company and its business partners has also brought with it the increasingly clear definition of interfaces between steps in a process inside a company and between business partners in a supply chain. In order to exchange data electronically between systems, the interfaces have to be made much clearer than in the case of communication between people. This is because computers never ask complementary questions, and communication simply breaks down if interfaces, messages and their interpretation are not well defined.

Increasingly, this development has brought with it the divestment of business units and functions, which are not seen by the company as core. This has given rise to a number of seemingly new businesses with a high level of specialization that work as, for example, third party logistics companies, computer service companies and security firms. To a large extent these functions were previously held by each large company itself. Through specialization and advantages of scale the services of these companies have become less expensive and they have also become available to many smaller companies that could, often, previously not afford to run these services on their own. Thus, the market for such specialized services has increased in size, specialized systems and other tools have been developed for these companies (which could not be justified when the scale of operations was small and limited to one single company) and the quality of services has improved.

This trend is usually called "outsourcing" and we have seen many new companies and industries arise from this. Outsourcing is a development toward a higher level of specialization both in the company that outsources a function and the company that offers the service. Many services with substantial volumes have now been outsourced by large firms. Noncore businesses have been divested, and interfaces internally in companies, and between business partners, have become defined and the web of interrelationships inside business has, to a large extent, become disentangled. This has also contributed to reduced cost and to the compression of time in business processes and supply chains. Now, specialized organisations with specialized tools cater to the specific needs of individuals and business customers all over the world.

From an initial situation of large-scale "chaos and disorder" we have now taken a number of giant leaps toward a relatively ordered and cost-effective situation on a global scale. It could be argued, however, that we have achieved cost-effectiveness on a global scale at the expense of pollution of the environment, climate change and resource depletion. With supply chains that require global transportation and ample and timely supply of oil and other energy sources, we now have to change to a more sustainable way of doing business. At the same time as the need for sustainable technologies, products, services and business practices is becoming overdue, we will start to face the exhaustion of the traditional competitive advantages of specialization, time compression and cost reduction. In many cases the opportunities to further increase specialization and reduce time and cost will be limited and the cost of making the next improvement will, at some point in the future, outweigh the benefits.

In order to find new areas for business development and future growth, we need to embrace sustainability as the third wave of business development, after strategic market dominance and operational efficiency. Energy transition, based on the drive for sustainability, represents a new area of business development and investments, and it also represents new opportunities for operational efficiency, as the initially relatively expensive sustainable technologies are made in larger volumes at reduced cost.

6.5.6   Summary of Operational Improvements and Energy

It should now be clear that the investments that have gone into BPR, lean management and supply chain management have been very large during the past two decades. These investments involve substantial investments in IT systems and collaboration, the complete redesign of vehicles, machinery and appliances and even the relocation of numerous key suppliers to locations closer to assembly plants. Through these efforts, which could serve as examples of large-scale structured attempts at organizational learning, the specialization of companies in supply chains has increased, and new industries have been developed, and existing industries have grown, through outsourcing and new services, such as consulting. These giant investments have been justified by the substantial cost reductions that have been made.

Buyers of these products have enjoyed reductions in price and we have, as a society, experienced increased levels of affluence and economic growth, due to the productivity improvements that have been achieved.

Cost reductions in the supply chains are usually proportional to time reductions and it should be obvious that the savings that companies have achieved in the past will not be repeated in the future, since companies normally work on the largest improvement opportunities first. Reducing the time from order to delivery from a few weeks down to hours, with the reduction of several weeks and a large amount of cost as a result, could not possibly be repeated. Nothing can be done in less than no time or at a cost that is lower than zero.

The changes to processes and supply chains that we have discussed above have also been taken under the assumption that low energy prices will remain low. We now have global supplies of clothes from Asia and other low-cost regions to more developed markets. These production structures have replaced existing structures, which were based on local or regional production and distribution. Garments are flown in or transported by boat and truck. These supply chains are costly in terms of energy, compared to chains a few decades ago, when volume items such as clothes and food were typically produced closer to the market, in many cases using raw materials that were available locally or regionally.

The changes in supply chains and in the way that companies work internally have not only improved cost efficiency and increased the need for transportation and energy in order to get necessary goods to the market. The development has also destroyed earlier local and regional supplier networks and replaced them with the global supply chains that we have described above. Even if some of the old companies still exist, it would, in many industries, take years of huge investments to reestablish more local and regional supply chains and distribution. This, to some extent, however, may be exactly what we need to do in the next few years, but before we decide which alternatives to choose, we need to investigate different alternatives, make a strategy and a plan for the change and start a program of planned implementation. As we have seen, there are different levels of cost, time frames and resource needs attached to different alternatives, and we need to understand the different alternatives from a change management–related perspective, before we will be able to incorporate them into a plan.

6.5.7   Specialization and Value Added

It has to be emphasized at this point that the model for economic growth that we have applied in the West is built on the concept of specialization and the pursuit of advantages of scale and learning in increasingly specialized businesses. It is probably no exaggeration to say that this is at present the only model for large-scale economic growth that we have thoroughly tested. Even though we do not, as has been admitted by economist Paul Krugman,[4] know very much about the exact mechanisms that create growth or how to manage this process at a high level to avoid recessions, this is the model that we collectively have the best understanding of.

In this model specialization leads to increasing value added per person in a value chain. Specialization has also made it possible to outsource low-value jobs to low-cost countries. This has historically contributed to our value creation in developed countries by making it possible for us to focus on tasks that add the most value and it has helped low-cost countries in their development, as is presently the case for Eastern Europe and large parts of Asia.

It may be possible to create large-scale economic growth by applying other models and it may, theoretically, be possible to transform global economies to stable state economies. This, however, would constitute a major challenge. We know how to solve technical problems even under immense time pressure and we know how to control technical risk and how to make cost reduction to work to spur economic growth. We also know how to rebuild economic systems or global supply chains almost from the ground up in a few decades, if this involves cost reductions and benefits in terms of strategic advantages for the companies that make the investments. We find it difficult today to envision a program that could be driven without the opportunity to achieve business benefits. This is not a pessimistic statement, but rather a statement of fact. The challenge that is involved in transforming economic systems, while maintaining a number of global virtuous economic circles, is, if we want to take it on, probably the biggest challenge that faces us in this transformation.

CHAPTER 7

Thoughts on Planning and Market Economics

7.1 THE MARKET ECONOMY

A market has the advantage of resources being allocated to the most efficient producers. The resource allocation "mechanism" is sometimes called the "invisible hand" of the market. This phrase was coined by British economist Adam Smith in the eighteenth century. Through the individual and free choices of a large number of independent consumers, resources are allocated to the different producers of consumer products and services that produce the highest-quality products at the lowest cost and who, in their turn, purchase the machinery, material and services that best contribute to their success in the consumer market. Through supply chains, back to the producers of raw materials, the most efficient suppliers at each level, from a number of different perspectives, are selected by customers, who provide the resources that fuel the growth of these suppliers. This, in some situations, is a very efficient process.

The market economy is very efficient for the allocation of resources, but the process is similar to that of the evolution processes in nature. The current car companies were established in the early decades of the twentieth century. Through the decades, they have learned to understand the preferences of their customers and how to develop and produce the cars that customers want. This process of learning to understand the market is a slow process and companies rarely want to, or need to, take large risks. When customer preferences, or other circumstances in a market, change rapidly, companies can no longer rely on their understanding of the market, and they sometimes lose large sums of money. When Japanese watch companies launched digital watches in the 1970s, the incumbent Swiss watch industry, with 80 percent global market share, was unable to adapt and the Swiss companies soon only had a meagre 20 percent of the global market.

Most companies find it difficult to even adapt to slow change. Peter Senge relates the example of the growth of Japanese cars in the US automobile market. US automakers did not react to this increasing competition in 1962, when the Japanese market share was below 4 percent. In 1967, it was less than 10 percent and the companies did not react. The same was true in 1974, when market share stood at less than 15 percent. In the early 1980s, the market share of the Japanese had risen to 21 percent and then the US producers started to react. In 2005, the market share was close to 40 percent, and the Japanese car companies, led by Toyota, have now gained the upper hand in the US market.[1]

Adaptation to rapid change is often very difficult and risky in a market economy. Companies that take big and decisive steps in situations of high uncertainty often get punished by the market and make big losses. The example of the development of the System 2000 by Philips in the video recorder industry in the 1980s is only one case in point. Philips decided to develop a new and technically superior video recording system. During the time it took to develop this technology and launch the new products in the market on a large scale, the VHS technology had already grown substantially. Soon, VHS became the de facto standard of this industry and the huge investment made by Philips was lost and a large new plant in the Netherlands for the production of these recorders lay idle for a number of years.

In most industries, companies have learned from these lessons, and they tend to take small steps in technology development and market investments, rather than huge risky leaps. In most cases, it takes at least 5 years to develop a new technology through adaptation of existing technologies, based on existing general concepts, and produce prototypes. According to Vernon W. Ruttan in *Is War Necessary for Economic Growth?* it takes decades to develop new general purpose technologies from scratch, and the investment necessary in order to achieve this cannot be borne by private companies. This is because of the volume of investment needed, the uncertainty and the long periods of consistent investment that are necessary before the volume growth starts to take place. After the development of prototypes, production is started at a low level, as the segment of customers usually known as "early adopters" starts to purchase the product at a high price. At this point, the market share of the new technology slowly grows. At the present point in time, the development of hybrid cars and cars run on biofuels has arrived at this stage.

If initial market tests prove to be favorable, companies gradually increase production capacity. The rate of this growth depends on the competition and the price of the novelty relative to competing products. In the cases of mobile phones and faxes, these innovations entered markets without any competing products in existence and grew rapidly. These new gadgets were also relatively inexpensive and could be afforded by a large share of their potential customers immediately, or within a few years. In the case of hybrid cars and other sustainable technologies, these often compete with an installed base of existing products that have technical lifetimes of two decades or more. Used cars are sometimes traded for decades and it takes 17 years until half of all cars from a particular year are taken off the streets. A majority of customers are used to the existing technologies and view the new technologies with some suspicion. Further, products using existing technologies are produced in large volumes at low cost. Customers who want to contribute to sustainability have to pay a large premium, but only a minority of all customers can afford, or are willing to do, this.

In the case of growth of the market for cars and other products that use renewable fuels, there is also the problem of the uncertainty about the volume of fuel that is available now and the fuel that will be available in the future. It is possible that many people would want to purchase cars that run on biological fuels, or that many utilities companies would want to build power plants that are fueled by other renewable fuels. As these technologies become more commonplace, we may reach the maximum level of production of these fuels. Limits may be set through restricted access to raw materials for the production of fuels or limited areas of land that can be used in order to grow energy crops. As we will see later in the book, the complex interrelationships between engine technologies, renewable fuels and distribution systems for these will, most probably, prove too knotty for the market itself to handle in a quick manner.

In very complex situations, where several players are involved in the development and marketing of a number of different products that are dependent on each other, decision makers will inevitably feel uncertain about one or several aspects of the future playing field. Usually, in such situations, development slows down and the technologies and products that could have been developed take a long time to get to the market, and the expansion of production capacity could also take a long time. In order to speed up development and the increase of production capacity in these areas, we will need a strategy and a plan for the overall change. This is currently the situation in the transportation sector where a number of different players need to make large-scale concerted efforts.

In other situations, as in the need to invest in new capacity for electricity generation overall, and the need to invest in renewable generation technologies in particular, the market signals in the form of customer demand and a willingness to pay the price of renewable energy will probably be too weak to initiate the high level of investment activity that will perhaps become necessary. While sometimes devastating, many of the consequences of climate change are not consistent and predictable enough to translate into market signals that could drive a particular investment wave in a particular industry, country or region forward. Legislation, taxes, subsidies or other change management measures will only be possible in the light of an overall view of the challenges, environmental concerns, time frame, cost, resources and investment needs. We need a strategy and a plan for the change.

7.1.1   Markets, Planning and Organizational Learning

Organizational learning is a field of research of growing importance. To a large extent the interest in this field has been triggered during the past few decades by experiences from change projects in companies. It has for a long time been noted that sending employees to training courses has little effect, in terms of change in behavior and routines at work, if only one or a few employees are sent at a time and if the training is not part of a change program. In the same way the launch of a new strategy by management in a company has little effect on behavior within the organization if the strategic change is not accompanied by a program for implementing the strategy. Such a program needs to include large-scale training efforts, and often also other types of support for those individuals in the organization who are supposed to change their behavior as a consequence of the new strategy. Companies that have experience in large-scale change programs, such as GE, know that any change effort requires much more resources than simple training and information activities. Change management is the process of managing organizational learning.

Despite these experiences that have primarily been made during the past decades, the concept of organizational learning was first, as mentioned above, conceptualized by Boston Consulting Group. This company discovered, through analysis of the development of production cost, that the cost of producing one unit of a product is reduced by the same percentage every time the accumulated volume of production doubles. As has been mentioned above, there is usually a cost reduction of 15–20 percent every time the accumulated production volume is doubled. The more high-tech the production process, the steeper the "learning curve". This means that the second unit produced is normally 15 percent less expensive than the first, and the 2000th unit is, similarly, 15 percent less expensive compared to the 1000th. This relationship is expected to continue forever. Even for beverage cans that are produced by the billions every year, the 200 billionth is expected to be 15 percent less expensive than the 100 billionth.

The process behind this development was called "organizational learning" and it was found that the cost reduction could not be pinned down to a simple relationship, such as the reduction in purchasing prices for supplies with increasing volumes or increased automation that may also become possible as a consequence of increasing volumes. Instead, the 15 percent reduction is curiously persistent, almost as a law of nature, and it has multiple causes:

- Improved routines for internal information processing
- Increased efficiency of production and distribution processes
- Cooperation with suppliers that leads to lower purchasing cost because of better management of the supply chain system and reduced cost of communication, deliveries, quality control and increased volume
- Increased automation of production and administration through mechanization, robotics and IT
- Standardization of components, systems and business processes
- Other causes

All of these examples boil down to the fact that individuals who work together in a system with similar goals, such as a company, government or society as a whole, learn how to gradually improve numerous aspects of their work, so that the whole system works better all the time. Due to the fact that the focus in companies and in society has been on cost and efficiency, which contributes to corporate profit and economic growth, organizational learning has to a large extent been focused on improving these factors. However, the focus on other factors, such as improvements of work life, environmental aspects of business and corporate social responsibility, has set in motion similar processes of organizational learning in these areas.

Progress in organizations, along all parameters, is gradual due to the large number of individuals that are involved and organizational learning tends be slower in larger organizations, and slower yet in society as a whole. However, GE, one of the world's largest companies, is evidence to the fact that a well-managed large company, despite its size, can learn to be almost as responsive as a small company to changing circumstances in its environment. As described later in this book, large companies have the advantage over small companies when it comes to change and responsiveness, due to its large financial resources, but the disadvantage that a large number of people need to be directly or indirectly involved in any effort to change. As rapid change has become a matter of survival of many large companies, they have taken on the challenge of managing change and organizational learning in their own organizations.

In society, organizational learning on the overall level has contributed to development and economic growth. On this level we have gradually learned to organize change activities with increasing efficiency, utilizing the market forces in an increasingly efficient manner (from a business and financial perspective) and use planning and managed change where this is more appropriate. Without doubt, the focus on economic value has led us to sacrifice other values in the process. The present belief in market economics can, from a historic perspective, be seen as a swing of the pendulum to an extreme, a reaction to a strong belief in planned economic growth that culminated with the Soviet and Chinese experiments. The very strong belief in the free market of the present has to be followed by a period of moderation and possibly another swing toward the planning extreme, which may be forced by the rougher energy and economic circumstances that we may face in the near future.

Organizational learning also has a not so well-known side effect. Organizational learning leads to increasingly complex organizations in which the specialization of employees gradually increases. This means that all employees perform an increasingly well-specified and narrow set of tasks. Organizational learning includes the process of sorting out tasks that add value to customers and reduce the time spent on tasks that do not add value. Automation and the implementation of IT systems also contribute to the increasingly narrow formulation of tasks.

This specialization makes organizations, and their employees, as well as society as a whole very responsive to challenges and change management needs that have been foreseen by management, but almost entirely blind to threats that are unexpected. It also means that managers, who carry the overall responsibility in organizations and should be the first to react to emerging threats, spend most of their time managing processes that are well known and foreseeable. Present-day managers have by now become highly skilled in their various areas of management, and their performance is measured in terms of their success in these areas. As a consequence of this, very few people in companies, or in government, spend their time studying the challenges of the future. The challenge of energy transformation had been better handled by the perhaps less efficient, but more generalist, managers of the 1950s, than by the increasingly myopic managers of today who are primarily driven by quarterly earnings than by long-term survival of our society.

Interestingly, this increasing myopia of managers, through the process of organizational learning, may be built into the very fabric of human society. The archaeologist Joseph Tainter has developed the theory of decline of civilizations that is currently accepted among scholars as the best explanation of the fall of ancient societies like the Maya, the Roman Empire and a large number of other less well-known societies of the past.[2] Tainter concludes that the demise of societies of the past is best explained by the fact that increasing complexity, measured as the increasing specialization of individual roles in society, made these societies less fit to cope with new challenges. In fact, Tainter explains that the fall of the Roman Empire was caused by the invasion of peoples from Asia, such as the Huns. The fall, however, was not caused by the invasion itself, because similar invasions had been successfully averted earlier in the development of the empire. Instead, the increasing complexity of Roman society, and the lack of large new opportunities for expansion (such as rich countries to invade), made the empire unfit to cope with the rising challenge. Lacking these opportunities to expand, the resource burden of complexity could no longer be sustained and the empire collapsed. In a similar way other ancient societies collapsed when they had to face challenges that they had been able to cope with earlier in their development, such as droughts and other changes in their environments. Specialization makes us more efficient short term, but unfit to cope with the unexpected. Energy systems transformations that we have previously been able to cope with many times over in the past, such as the transitions from whale oil to gas to electricity for city lighting,[3] may become an increasingly tough challenge for a more complex society.

7.1.2   Planned Decision Making in Market Economies

As we have seen above, the US economy was turned into a war economy in less than 1 year, and within 3.5 years of this decision, hundreds of thousands of airplanes, tanks, ships and other vessels and vehicles had been produced. During the same period of time, millions of uniforms had been sewn and large numbers of boots, rifles and other necessary materials had been produced as well. In the case of the Apollo program, after President John F. Kennedy's decision to put a man on the Moon, it took only 8 years to accomplish the mission. During this time a host of new technologies had been developed and integrated with each other into functioning spacecraft; a number of test flights had been performed, using, first, smaller craft, and, later, full-scale *Saturn V* rockets. Large-scale launch facilities had been built in Florida, and a control center had been erected in Houston. Around 400,000 people, primarily in the US, had participated in the program in some way, and 20,000 suppliers had been involved in supplying technologies, material or services to the space program. As recently as in 1958 the Space Task Group was formed, starting with 45 persons.

Such rapid transitions and development efforts would have been impossible to achieve through market-based resource allocation. Few companies would have dared to take the risk of developing entirely new types of aircraft for the war or escalating production resources of them in the way that the US government did in 1941 and 1942.

True, products in this case may have some shortcomings, due to lack of time to patiently test them on a small scale, and budgets may be overdrawn. Customers will not have the time to react to the first series and collectively make an impact on the development of the second generation of planes 2 years later. The same is true for the Apollo program. There had to be a government agency, NASA, that decided which technologies had to be developed for the Apollo program. This agency had to evaluate different alternatives and proposals, and then, at a relatively early stage in the program, decide which technologies to use and also decide about the interfaces between these technologies. Also, often, in this type of situation budgets may be large and there may be insufficient pressure on companies to keep tabs on cost. However, similar to the purchasing strategy during the Marshall Plan and the purchasing of material for the war, efforts could be made to maintain market-based interaction in as many situations as possible. In a large-scale expansion of the production of energy-efficient products of various kinds, the rapid increase in volumes will provide advantages of scale and a rapid descent along the learning curve in terms of cost, quality and performance. This will add financial benefits that we will be able to rapidly take advantage of in the program.

The advantage of planned decision making is speed. Speed is achieved by lowering the risk for companies and investors. In this case, the government would accept taking the risk of each project that is included in the overall program. The government may, in one type of scenario, place orders and, perhaps, pay for the development and guarantee the purchase of a certain number of the product. In other cases, such as in the case of the Marshall Plan, the government supports a certain development through direct investment, loans or making other obligations.

As has already been argued, this book does not propose a planned economy. It offers three examples of planned programs and three different ways of putting plans into practice. It is for the strategy development phase of the overall energy transition program to contribute a road map for the change and suggest a particular mix of market mechanisms and centralized planning. The complexity of many of the markets for sustainable technologies and fuels that need to be created suggests that some level of planning is needed. The cry for rapid change that is raised by a growing number of experts suggests that the level of planning needs to be relatively high. Planning is the way to make sure that goals are achieved.

Planning, however, does not necessarily imply that the government will have to pay for everything. It only means that it makes sure that all important aspects of change are planned for and taken care of by some player in the economy. In case it finds that a particular aspect of change will not be taken care of, regardless of the reason, the government may take certain steps to make sure that this aspect is covered in a satisfactory way. In a market there is nobody who can be held accountable for achieving the goals set up by the government, not even the people who set them in the first place. This is another not-so-subtle difference between a market and a planned program and the reason why any pressing program will need to be accompanied by a plan.

7.2 PLANNING – A WAY OF MANAGING RISK

It is argued above that, for companies, planning is a way of reducing uncertainty and risk. By putting in place budgeting and other planning procedures and by allocating resources to different activities and development projects, corporate management can control the risk level in the company and keep as much control as possible over overall development. This means that the risk exposure, which is, among other things, created by running risky development projects, could be balanced by foreseeable revenues and profits. Sometimes the cost of a project that has failed has to be written off, but this should happen as seldom as possible and risks need to be taken under highly controlled circumstances.

Companies rarely undertake very large and risky endeavors that could cause large-scale losses and possibly put the future of the whole company at peril. Development of new technologies and products are generally risky projects. Small-scale development of a new technology, feasibility studies and the development of prototypes are not very costly or risky for large corporations and such projects could be seen as necessary development activities that need to be performed in order to get any knowledge about the technologies that are promising for large-scale applications.

The same is true for other types of projects, such as projects that would be related to the transformation of supply chains and create local and regional suppliers. Without a clear and foreseeable financial upside for this type of endeavor, it would be impossible for companies to follow this path and finance it from their own funds. Even if the move were neutral from a cost perspective, it may not necessarily be made because of the management imperative of investing money in profitable activities.

We need to expect that companies, and the financial markets, will be reluctant to make large volumes of financing available for unprofitable or very risky endeavors. This is not because the people who work in finance could not be made aware of the pressing environmental needs that our society is facing. Surely, many are already painfully aware of them. The remit of a manager, however, is to invest the money of a company in projects and opportunities that will return a profit. With our present situation, there are no mechanisms in place that could, on a large scale, turn companies into charity organizations that make large investments in unprofitable projects. If such investments are needed, planned action will have to be taken and this needs to be based on an understanding of the financial and competitive consequences of such measures.

7.3   MARKET MECHANISMS AND MARKET IMPERFECTIONS

In theory, the market mechanism is able to immediately allocate resources to projects and investment opportunities that offer the prospect of positive payback. In reality, this mechanism is not immediate. The limited knowledge of opportunities to save money and make profits on investments in the application of energy-efficient technology, or on the development and commercialization of this technology, needs to be taken into account, and this imperfection, in terms of knowledge, accounts for substantial delays in the development and application of new technology. Another source of imperfection is the delay from the time a market opportunity is identified, until the time a solution is developed and implemented and has penetrated the market. Thus, even if companies and consumers today find that it makes financial sense to invest in a car with a new engine technology, the production resources for vehicles using this technology may not be developed to the point where a large share of customers in the market could be supplied with the product. The rapid reallocation of resources, which is possible in financial markets, will not be equally rapid in markets for vehicles and other technical equipment. Another factor that limits the speed of resource reallocation is the fact that vehicles, houses and plants have economic and technical life span of decades. Again it takes 17 years to replace half of the vehicles made in a year by new vehicles.[4]

First, not many people in business are aware of all the opportunities that exist for saving money by using less energy, or of the opportunities that exist to develop products that save energy. Many people are just now awakening to these opportunities that are very diverse. For any company that uses energy, it will take a substantial amount of time to analyze the savings potential of all the available opportunities, using existing sustainable technology alternatives. There are a multitude of activities in a company that consume energy in one way or another. In some cases it is relatively straightforward to analyze these activities and evaluate the cost and savings potential of different solutions. This is the case of industrial installations, such as pumps, lighting and heating. In the case of a technology used in production, such as pneumatics, it is easy to calculate the cost of existing installations, but there may be few solutions known that improve the energy efficiency of pneumatics. At present, as an example, not a lot of people will be aware of X-Block (described elsewhere in this book). In the case of business travel, it may also be easy to calculate the energy that is expended on a certain business trip, and the cost and savings potential of reducing business travel. In this case, the consequences of this type of measure may be up for debate, and it may not be clear which trips are necessary from a business perspective, and which activities can be handled without travel. We could collect best practices from energy-conscious companies, but the clear-cut calculations of financially justifiable energy savings, which are taken by economists to exist and that could form the basis of perfectly rational business decisions, cannot often be produced.

The case becomes even more complicated by the fact that new technology is currently developed and launched in the market at a very high pace. The solutions that are available right now may be outdated in a few months. No company can keep track of all opportunities and constantly evaluate every new technology or alternative business practice. As new technologies, and new products that make use of existing and improved technologies, become available, it will take time until all potential customers become aware of the savings potential offered by them. Furthermore, many customers may wait to make decisions until the next generation of technology is available or until the market has stabilized and a standard has been developed. This uncertainty and avoidance of making quick decisions will in itself delay the development of a standard and stall the whole development.

Second, in the same way, it is impossible for investors, who evaluate the opportunity of investing in a new energy-efficient technology, to estimate the exact market potential of this technology. This creates two types of risk. One is the risk that investors for a long time underestimate the market potential of energy-efficient technologies. When GE launched the products that were developed within its project Ecomagination, this company was surprised at the unexpectedly high demand that these technologies now generate in this market. It is probable that many investors have been underestimating the potential of such products, because relatively little has been written about business opportunities in this area. In this case, the market will not allocate resources, even to perfectly justifiable product development opportunities, until the expectations of investors become more realistic.

Unfortunately, however, investor expectations are seldom realistic, or at least they are unrealistic often enough to make a point of this. When investors suddenly become aware of the opportunities that open up in the field of green products and services, it is probable that they will overestimate these opportunities, or perhaps underestimate the number of other investors that try to pursue the same opportunities. These two factors may work in combination. At some point in the development, a large number of investors overestimate the opportunities themselves, believing that the market for energy-efficient products and services will grow faster and become more profitable, than will really be the case. This will lead to the "boom and bust" cycles that characterize markets and it is a relatively resource-inefficient way of running development, especially in a situation when rapid and reliable progress is needed. A certain measure of overall planning on the part of society could add speed and reduce the cost of this endeavor.

These things have happened repeatedly in the past, as investors have created speculation bubbles. During the Internet boom, investors invested huge sums of money in Internet-related stock, even though many were aware that this had to be a speculation bubble. At the same time, many investors pursued similar business opportunities, so that the number of Internet sites that were developed for trade in steel, or transportation space could be counted by the dozen. During the shake-out substantial sums of money were lost, and for a number of years, the interest even in sound Internet-related investments waned. The long-term and slow periods of small-scale trial of concepts and product and service ideas, which are necessary in a market economy, were not gone through. Instead investors and managers with little experience from the developing market made unrealistic assumptions and many simple mistakes.

In the case of the energy transition, we have both limited time and limited financial resources. We need to rapidly expand the use of existing energy-efficient technologies and products, and develop a number of new ones in areas where we presently lack sufficient solutions or production resources, such as in transportation. It is critical that we do not waste time or resources, or deplete the confidence of investors and buyers, in these technologies. This is another argument to the advantage of planning.

Box 7.1 A Brief History of Economic Growth Theories

The history of economic theory indicates a gradually growing understanding of economic growth and also a growing understanding of the reasons behind our improved affluence. Still, economics professor Paul Krugman, in his book *Peddling Prosperity*, argues that economists still do not know enough about economic growth in order to, in each situation, be able to develop particular economic or political programs that will lead to economic growth. He argues that the economists behind political strategies, such as "Reaganomics" or the development programs of Bill Clinton, are not founded on economic science. Contrary to popular belief a single president during his presidency can actually influence economic growth to a very small degree. Factors other than the policies of the president are more important, such as the unemployment rate at the time when a president took office. If unemployment is high, many people can be put to work, which leads to economic growth. Other factors, such as the global business cycle, are often more important than the economic policy.[5]

Over the longer term, however, the actual configuration of an economy is important for growth, and the understanding of how economic growth is created has developed through a number of significant theoretical steps. The first important theory of economics, that we still use today, is the theory of the free market, and the power of the "invisible hand" of the market to allocate resources to the most efficient producers in such a market. This theory was introduced in 1776 by British economist Adam Smith. According to this point of view, the role of the government is supposed to be limited to legislation, and the government should try to interfere as little as possible in the workings of the market. At that time, markets were not very complicated, and the typical example of a market would be a farmer's market for grain. In order for a market to work perfectly, all producers and customers will need perfect information about supply and demand, and also perfect information about the production resources, and the different costs of production. Any limitations in the access to information, or the inability of players in the market to gather, or analyze, information, will be seen as imperfections in the market. Imperfections will cause time delays for product development, capacity expansion and other necessary activities. This gives the theory of the invisible hand of the market-limited applicability in the real world. Still, there is a strong belief in this theory, and little understanding that market imperfections need to be analyzed and understood.

Many economists hold that British economist John Maynard Keynes was the most important economist of the twentieth century. Keynes argued that governments should try to control economic growth through fiscal policy. During a recession, governments should increase spending, in order to create new jobs and stimulate the economy to grow. The opposite should be done during an economic boom, when the government needs to exercise restraint, in order to reduce demand for labor and other resources.

In his work *Capitalism and Freedom* Milton Friedman introduced the concept of "monetarism," which means that the government and national banks, should control economic growth by controlling the money supply in the economy. He argued that the Keynesian policies of governments came too late to be effective. A stimulation package tends to be discussed and planned for so long that when it is finally implemented, in most cases, decline has already turned into growth, and the package is likely to add to the instability, rather than function as a counterweight. Efforts to control the money supply would have a much quicker effect on economic development than Keynesian policies, and due to short lead times, they would mostly have the intended effect.

In the 1980s the work of Mancur Olsen highlighted the effects of collective action and institutions on economic growth and prosperity and Olsen also used his models to analyze the rise and decline of nations. The impact of institutions on economic growth is still an important research topic in economics.[6]

During the 1990s most economists agreed that the most important factor behind growth is the improvement of productivity.[7] Productivity improvements, such as when companies automate production or increase their use of computers in administration, free resources that can be used in the economy to produce more goods and services. This realization, the fall of the Soviet Union and the apparent success of "Reaganomics" and "Thatcherism" contributed to the growing belief in the free market, privatization and deregulation. Even though some economists maintained an interest in factors other than free market economics, the public debate and the bulk of economic research dealt with different aspects of market economics.

In the first years of the new millennium, Professor Vernon W. Ruttan published two books that once more emphasized the importance of government investments and expenditure in economic theory. In the way that has already been mentioned in this book, he found that government projects over the past centuries have substantially contributed to the development of "general purpose technologies," the application of which by companies and by society at large drives economic growth. The development of technologies, such as aviation, nuclear power, computers, space technologies and the Internet, requires very large long-term investments that cannot be mustered by private companies that operate under market conditions. Instead it is the continuous investment, during decades, in development projects that make the development of this type of technology possible. Contrary to Keynes, Ruttan does not say that governments should influence economic growth short term, by increasing or decreasing spending. He argues that governments need to invest continuously over sustained periods of time, in order to develop the technologies that will renew the economy in the future.

Government-funded technologies have, on a large scale, contributed to the "creative destruction"[8] in economies in the past, through which Atlantic liners and transcontinental train travel were replaced by the growth of airlines, and as the growth of the Internet is now replacing traditional music distribution and the distribution of printed matter by electronic communication.

Throughout this development of economic theory, economists have described the importance of different mechanisms in the economy, and how different mechanisms contribute to economic growth and affluence. Despite the fact that a mix of different factors has been present in the economy for a very long time, such as government-funded technology development, which, according to Ruttan, goes back at least to the early nineteenth century, it has taken time to understand the importance of such factors. New findings, such as the understanding of the importance of general purpose technologies, open up new areas of understanding, and new opportunities and needs for economic policy. We need to assume that this development will continue into the future, and that there are, and will be, more mechanisms and tools available to policy makers, than the ones that have been identified and described by economists so far.

Among the areas that are now being analyzed are the ones related to complementary currencies and stable state economics. In stable state economics the imperative of economic growth, which is part of our existing economic system, has been removed. This is achieved by installing different mechanisms into economic and monetary systems than the mechanisms of interest rates and expansive economic policy. The advocates of stable state economics sometimes also advocate complementary currency systems that are described elsewhere in this book, under possible financial tools for the transformation. These ideas could to a certain extent complement existing economic solutions and should not be rejected offhand because they have not yet been tried on a large scale.

7.4 LIMITS IN TERMS OF TIME AND OTHER RESOURCES

In any company at any time, there will always be a number of different investment opportunities and projects that can be run. These range from the development of new products, to operational improvements and improvements of customer service. In a pure market economy, any company will have to prioritize its time and other resources, and pursue the opportunities that make the most sense from a financial perspective. A company may identify 10 or 20 different energy improvement opportunities that would be financially justifiable. In competition with other projects, such as product development opportunities or other investments in expansion, these opportunities may still receive a low priority and be put on hold, while a number of other projects are run. The impossibility of pursuing all financially justifiable opportunities at once represents another market imperfection and a difference between the real world and the way that economic theory is interpreted by many politicians and other key people in our economies.

There is a need for planning at the level of society to identify the most important improvements in companies and probably make some of the financial and other resources that are needed by companies available at low cost or make it mandatory, by law, for companies to make certain improvements in terms of energy efficiency.

Even the emission rights that are now traded, as part of the Kyoto Protocol, may not have the effect that the governments expect them to have. Many companies may wait to make improvements at the last moment, and try to purchase emission rights at this point, when they realize that they will have too little time to act. Major improvement projects will take several months or years to go through, and companies that do not start early enough may simply fail to improve rapidly enough. This may lead to a situation, in which more emission rights are in demand than the numbers that are available in the market. True, the price of rights will go up, but some companies that fail to acquire rights may still not be able to make the necessary improvements in a very short period of time, and the savings goals of the program may not be met.

7.5   NATURE AND MARKET MECHANISMS

As already mentioned, at the time when the price of energy reaches the level when an alternative technology will be financially justifiable, it will take several years to develop technologies and products that could replace the old technologies. In case a sustainable technology and a number of products are already in existence, it will, in many cases, take years to expand production resources so as to make supply meet demand.

Nature, however, does not care about market mechanisms. The price of a new technology will not necessarily reflect the necessity of the technology from the perspective of nature or the climate. The application of sustainable technologies may be long overdue, but the price of fossil fuels or other incumbent technologies may still not be high enough to justify widespread investment in new technology. It is possible that many companies spend so little on energy that the issue of energy savings will not get a high enough priority. Some 70 percent of the energy used in business is used by a very large number of small users, who use small amounts for a number of different activities in production, transportation and travel. Even if the need of reducing emissions of greenhouse gases is high, the savings potential for many companies will not be very high.

In such cases, there will also be a need for planning, so that companies that do not see saving energy as an important activity will be guided and supported in their efforts to save energy, or, perhaps, forced to do so through legislation. This book does not argue in favor of any particular solution, other than making planning and, possibly, government funding two tools that need to be included in the toolbox that will drive energy transformation.

The market may be a potent tool for long-term economic growth, but we need to understand that its mechanisms are not precise enough in order to ascertain a certain outcome at a particular and foreseeable point in time. As I have argued above, there are imperfections in all markets, related to lack of information and resources for analysis, and delays in market response, which in some cases will make the market a poor, or insufficient, tool for the achievement of particular goals. As has already been mentioned above, foreseeable increases in the price of energy will be much clearer signals to companies, but even these will neither come early enough nor be clear or strong enough to initiate all necessary energy improvement activities.

Part II

Organizational Learning and Change Management

As has been indicated above, change in complex social systems is a matter of organizational learning and change management. This part of the book describes a number of key concepts related to these subjects.

TEN STEP LISTS AS GUIDELINES FOR DECISION MAKERS IN BUSINESS AND GOVERNMENT

As a response to a question from Julian Darley at the Post Carbon Institute, I wrote two "ten step lists" that could function as guidelines for decision makers. The lists turned out to form relevant summaries of Part II of this book from a business and government perspective. As organizational learning will gradually change organizations and society and adds energy transformation competence in most organizations, it will become increasingly clear and well defined what this means in particular business situations and industries. At the time of writing, only a few readers will be in a position to act on these issues, but the number of people who will be able to do so will increase over time.

TEN ACTIVITIES FOR BUSINESS MANAGERS

1. **Understand** how increasing energy prices will impact the business
2. **Develop** scenarios for energy price and supply
3. **Analyze** consequences of different scenarios for growth and profitability and analyze alternative mitigation activities
4. **Prioritize** and plan energy transformation activities
   - Improvement of own products and services based on life cycle analysis
   - Opportunities to improve internally
   - Reduction in energy for transportation

5. **Mitigate:** Plan and prioritize mitigation activities
6. **Train** key decision makers at all levels in energy transformation management
7. **Lobby** for government action to organize activities, set goals and standards and run coordination projects in areas that will not be transformed through market-based action
8. **Work** together with suppliers to reduce energy consumption in supply chains
9. **Implement** a company-wide project and investment plan for improvement based on best practices
10. **Monitor** progress and manage the change process according to a tight transformation plan

TEN ACTIVITIES FOR POLITICIANS, GOVERNMENT ADMINISTRATORS AND ECONOMIC POLICY MAKERS

1. **Understand** the situation and key implications for the future
2. **Develop** scenarios for energy price and supply development
3. **Analyze** consequences of different scenarios for society and for the economy
4. **Prioritize** and plan energy transformation activities
5. **Foresee** (specifically) consequences for national and global economic systems
6. **Mitigate:** Plan and prioritize mitigation activities related to these systems
7. **Train** key decision makers at all levels in energy transformation management
8. **Inform** general public and small business owners about the need to reduce energy consumption and possible behavior and work life changes to facilitate this
9. **Implement** large-scale project and investment plans based on experiences from previous national and global large-scale projects
10. **Monitor** progress and manage the change process according to a tight transformation plan

CHAPTER 8

Two Levels of Control – Executive Direction and Program and Project Management

The main argument of this book is that four steps will be necessary in order to run a focused and speedy energy transformation on a global scale. These steps are analysis, strategy development, planning and managed change. The set of tools that will be necessary in order to accomplish this change will contain a large number of tools that will have to be used by steering committees, managers and program and project participants. A selection of tools for different purposes will be described in the following chapters.

Throughout the book, three terms are used to differentiate between the different levels of organization in the energy transition. For the overall level of the whole transition, the term "program" is used. The intermediate high level is called "stream." In this book six "streams" are identified, namely the streams of transportation, industry, housing, utilities, agriculture and behavior and work life change. Within each stream, there will be a number of "projects," in which technology is developed or other activities are undertaken, that are necessary for the success of the stream. In an actual change program of this magnitude, more levels will be needed, but for the sake of discussion in this book, these three will suffice.

In any change effort, whether it is a program, stream or project, there will be two overall levels of control that need to be taken care of. I call the top level of control, which is often handled by a steering committee, or a management team of a company, "executive direction." The day-to-day management activities in a project will be called "management," which represents the second level of control in the hierarchy of terms.

Executive direction is related to the strategic level of program, stream and project control. The steering committee has a role that is similar to that of the board of a company, making decisions on overall issues, strategy, budget, goals and plans. Setting guidelines for communication between the project in question and other projects is also a matter of executive direction.

8.1   THE OVERALL DECISION – THE COMBINATION OF MARKET-BASED AND PLANNED CHANGE

Overall, in the development of society and companies, there are two tools that are used in combination. One of them is the market and the mechanisms that form a market economy. The other is planning, which is used in many more instances than in a planned or socialist economy. In a market economy, companies themselves represent planned economies, which compete against each other in the market. The companies that are most successful at the planned development of its competitive advantages, gain market share and improve profitability, while the companies that are less adept at organization and planning lose market share and their profitability will be reduced. Referring to the discussion in the previous chapter, we will need to combine market mechanisms with planned change efforts in order to achieve the expected result on time.

8.2   THE TOOLS OF A STEERING COMMITTEE

In order to be able to take on the responsibility of steering the important projects within a stream, or in order to take on the steering of the program or the streams themselves, the steering committee, or the management teams that may be appointed by the government agency that becomes responsible for managing the program, will need access to a number of tools and resources for this task. In the case of transformation projects in companies or at lower levels in communities or regions, I still think in terms of a steering committee and the tools presented here are relevant for use in these situations as well. The individuals of the steering committee will not, in most cases, work with the tools, themselves. They are more likely to assign the completion of analyses and plans to program or project management or to consultants, and oversee the process of analysis, strategy development and planning performed by project management. The steering committee will also decide about strategies, plans and other documents, once they have been completed by managers. The following would constitute the key components of a steering committee tool kit:

1. *Visions, values and mental models*
2. A tool *for the positioning of projects* in relation to each other in terms of complexity, risk and pace.
3. *Guidelines for strategic priorities* in a project, given by the steering committee on the next higher level.

8.3   THE TOOLS OF PROGRAM OR PROJECT MANAGEMENT

As we move from the level of visions and values, we need to realize that a change process that needs to encompass the whole of society is, by definition, complex. In order for all streams and projects within a program to be managed in the same way, according to the same principles and values, a set of tools for program and project management needs to be put in place and used by project managers and participants. Such a set of tools will reduce the risk of individual idiosyncrasies of managers, or differences in administrative competence and drive, impacting the result of projects.

We have identified a number of necessary components of a change management program:

1. Overall *goals* that need to be determined for the project and broken down for lower levels of project management within the same project.
2. A change *strategy*, which determines the financial and other types of resources that will be needed, for example, people, technologies and production resources, and specifies the main steps in the change process and the method for driving the process.
3. A *plan* for the change program on the highest level and plans for change at lower levels, which are derived from the overall plan.
4. A program *organization* for managing and driving the change program needs to be designed and implemented. The people that are needed to fill the roles in the organization will have to be recruited or transferred from other jobs, if they are already available somewhere in the organization that is going to run the transformation.
5. *Management principles and guidelines* need to be put in place. This includes a meeting structure, a predetermined agenda for meetings at different levels and decision structures and a definition of which matters should be decided by a manager at a certain level.
6. *Methods and tools* that are going to be used need to be selected, developed or modified. There is also a need for checklists, for managers and project participants at different levels, budget forms and other tools. In this book, special attention is paid to tools for problem solving and planning as means to facilitate organizational learning.
7. A *communication strategy and plan* for the program need to be determined. Who is going to need information, at what intervals do people need to be informed and which information is going to be communicated?

8. *Training* of all participants in the program in all the skills that are necessary for completing their tasks.

9. *Change management* through the whole process.

Each of the above factors will be discussed from a theoretical and a practical perspective in Chapters 9 and 10. In Chapter 11 a number of tools for organizational learning will be described.

CHAPTER 9

Tools for Program and Project Executive Direction

9.1 VISIONS, VALUES AND MENTAL MODELS
(Item 1 on the list of tools for the Steering Committee)

In his management classic *The Fifth Discipline*, Peter Senge argues that visions, values and mental models will become the management tools of the future, replacing old tools, such as control and coordination. In the learning organizations, envisioned by Peter Senge, employees become empowered to make their own decisions, based on insight and an understanding of the situation. This line of thought represents a paradigm shift, compared to past thinking of management theorists and practitioners. Senge argues that employees at all levels should be seen as both willing and able to think for themselves and make their own decisions for the good of the company.[1]

The control paradigm of the past existed, we may assume, because knowledge, in those days, was less evenly spread across the organization, than at present. With a small number of managers at the top, who sometimes held university degrees, and who had information about the environment of the organization, it was less likely than today that people at lower levels would make relevant decisions. It was not until after the Second World War that higher education and business experience became more widespread, and decision making could gradually be dispersed more widely in an organization. This process has taken some time and is still under way.

In today's world, many employees resent the detailed instructions and control of the past, and embrace the opportunity to take a wider responsibility for the performance of their tasks. Hence, the role of management is gradually changing toward developing visions, spreading values and providing mental models for their employees. Leading, in the modern organization, becomes increasingly a matter of collecting signals from the organization, interpreting them and transforming them into coherent visions that could guide the direction of development of the organization. Based on the visions, managers, together with employees, need to formulate the values, which are needed in order to drive development in the right direction. Values could be of two basic types, namely, those that only offer guidance for a certain type of behavior and those that are firmly based in an ethical belief.

Box 9.1 Two types of values

My colleague Kjell Persson has identified two general types of values that are used to guide decision making in organizations:

1. Values that guide the behavior of employees
2. Values that find their basis in ethics

The first type of value is found in many companies in their instructions to employees. Many retailing firms nowadays have the policy of taking back any product that a consumer is dissatisfied with, no questions asked. The furniture retailer IKEA wants its employees always to think and behave in a cost-effective manner. Whenever possible, employees are supposed to save a few dollars on expenses. Even when there is no clear instruction in existence that could guide an employee at IKEA, the value is so strong in the organization that it provides guidance in all kinds of situations. A higher-level value could be represented by the GE value of being innovative and to go for new and innovative solutions, rather than following the crowd. Both these values from IKEA and GE have been implemented in the organizations by the founders, and then maintained and strengthened by later generations of top managers.

The Body Shop represents an example of ethics-based values. This company would never sell a product that has been tested on animals. In many cases this company also provides a living for native peoples and farmers in less developed countries, by purchasing their produce as raw materials for cosmetics. In addition to this, The Body Shop also encourages employees in shops and offices to participate in community projects in the cities where the company does business.

In both these types of cases, the values provide important guidance to employees in all kinds of business situations. For some situations, there may be written instructions, but for others, employees are supposed to draw their own conclusions about decisions, based on corporate values. In a business environment that is continually changing at an ever faster pace, values become increasingly important, because no company can foresee all the detailed business situations that its employees will encounter, and write detailed instructions for each of them. Instead, for intelligent, well-trained and well-informed employees, values can serve the same purpose.

In many organizations, the values held by an organization are seldom put into practice. If people get measured by the achievement of certain quantitative goals, actually living up to a set of values may be felt to be counterproductive. For instance, if a manager is measured by the goal of achieving maximum revenue and profit, it may be seen as counterproductive to take goods back that have already been sold to customers. In the influential management book by Tom Peters and Robert Waterman *In Search of Excellence*, which was published in 1981, the authors identified attention to customers as one trait that is related to excellence, but the authors also recognized that this, at the time, was an unusual trait in companies. What they did recognize was that companies that made an effort to provide better service to customers tended to become more profitable and grow faster than their competitors. In a later book *Built to Last*, James Collins and Jerry Porras identified "cultlike cultures" as one of the traits of the companies that showed sustainable growth and profitability, and they recognized that successful entrepreneurs are "clock builders" rather than "time tellers," meaning that the clock builders create organizations that work in the way that their founders envisioned, even after the founders left the organizations. Often they achieve this by instilling strong values in their employees, which they are expected to live by. In many successful companies, such as the electronics retailer Nordstrom, parts of the cultlike cultures are built on excellent service to customers.

In these cases, one way to create a culture that sustains a number of actions that may seem to be counterproductive over the short term is to implement a number of values, and support them through firm rules that cannot be misinterpreted. This could lead to a very large number of detailed rules. At Nordstrom, however, the entire employee handbook consists of a single card stating (emphasis in bold in original): "Welcome to Nordstrom. We're glad to have you with our Company. Our number one goal is to provide **outstanding customer service.** Set both your personal and professional goals high. We have great confidence in your ability to achieve them. Nordstrom Rules: Rule number 1: **Use your good judgement in all situations.** There will be no additional rules. Please feel free to ask your department manager, store manager or division general manager any question at any time."[2]

Values are sometimes developed and implemented by the founder of a company, such as in the cases described above, but they also need to be revised and updated, as circumstances in the business environment change. In an organization, they need to be constantly discussed. What does it mean, for instance, that employees at IKEA are supposed to save a penny on cost, whenever possible? Does this mean that managers on business trips should walk, instead of taking a taxi? Are they supposed to go by bus or by underground, instead of taking the fastest route between two points and save valuable time? In some cases, answers are obvious, in other cases they are less so. As companies enter new markets, mental models of how values are to be used may need to be updated. In a developed country, it may be appropriate for employees of a cost-conscious company to travel second class on trains. As such a company starts to do business in less developed markets, this may, for reasons of safety, health or convenience, need to change. While the values provide general guidance in many situations, mental models of how to behave in particular situations may arise as values are interpreted and discussed.

In the case of GE, the long-standing value of driving innovation, which goes back to founder Thomas Edison, has been revitalized by recent managers, such as Jack Welch and Jeffrey Immelt. New values, such as that of developing environmentally friendly products for a world that is increasingly in need of this, have been added more recently, derived both from the imperative of innovation and from the growing market need that this company has identified. This is an example of how values are kept up to date and alive in a large global organization and how one of the world's most innovative companies can constantly change, and yet always maintain some of its core values. As values change, new mental models arise, and replace the old and outdated ones.

9.1.1 Visions, Values and Mental Models in the Energy Transformation Program

The overall vision of the energy transformation program will be to reduce energy consumption and CO_2 emissions. In this book, the aspect of cost-effectiveness and time are introduced into the equation, two factors which have largely been absent in the debate up until now. Many technologies and opportunities with long lead times and alternatives that are still characterized by a substantial need for development are discussed in parallel with alternatives such as wind power, cogeneration of heat and power, hybrid engines and biofuels that are already available in small, but growing quantities. Even if the already available alternatives are not able to cover all of our future energy needs, we need to distinguish between different alternatives based on the readiness aspect and a number of other aspects when we make a strategy and plan for the future. This means that, in addition to the value of saving energy and reducing CO_2 emissions, values related to investment levels and payback time on the investment need to be developed.

This overall vision of emissions reduction and cost efficiency will need to be broken down into visions for concrete technology development projects. If it is decided that hybrid drivelines will need to become a common feature of trucks, buses and automobiles in the future, this needs to be formulated into a more concrete vision, such as one stating the percentage of hybrid vehicles that need to be in use by a certain year or date in the future.

This, in its turn, needs to be translated into technology choices and into values and mental models for program and project participants. How important is it to develop cost-effective solutions, compared to technically outstanding ones? In the case of the Apollo program, the goal was to develop one rocket technology, which could be used in a sequence of spacecraft, which could safely take astronauts to the Moon and bring them back. The emphasis, by necessity, was on technical excellence, rather than on cost-effectiveness. In the case of sustainable engine technologies and renewable fuels, these technologies will power vehicles for decades and the consequences of too costly solutions will influence the profitability and cost position of companies beyond the lifetime of the engineers that develop the new technologies. Therefore, values need to focus both on technical and design excellence to some degree, and on cost-effectiveness. Furthermore, the mental models of managers, engineers and sales people need to be focused on achieving high sales and quick market penetration of the products.

The steering committee of the program and of projects within the program needs to keep these goals in mind and provide visions and help in the formulation of values, together with project management and employees, which can help drive the program toward the achievement of overall goals. How pressing must the deadlines for technology, and product development, be? How rapidly must production resources be built in order to achieve sufficient market penetration, to reduce energy consumption and CO_2 emissions at the rate that is needed, or faster? What are the visions for new technology development and economic growth? In which way is the government, or the steering committees of the program, supposed to plan the development and assign roles to participating companies? Should government tenders be made official for competitive bidding by companies that have the right skills and resources, or should government agencies deliberately select the companies where they want to place development orders and production orders? It is probably the first alternative, as has been the case throughout the space program. Should government finance different parts of the program through direct investment, through the promise to purchase a certain number of products, or through subsidies or loans to buyers of the new technologies? Each such decision could give rise to a value or a set of values, but decisions could also be built on overall values that could be based on the relationship between market mechanisms and planned development in the program. In its turn, such a decision could be based on an idea of the type of business environment that top decision makers would like to create through the program.

9.2　A TOOL FOR THE POSITIONING OF SAVINGS OPPORTUNITIES (Item 2 on the list of tools for the Steering Committee)

Different opportunities need to be prioritized based on how much value is created through the use of energy and the cost of activities that are necessary to reduce the use on a large scale. Neither value nor cost is clear-cut. They need to be estimated area by area, and transformation activities need to be planned based on these estimates.

The matrix above provides a framework for this type of prioritization. Along the value axis the value of the use of energy for a certain purpose is rated. The value may be rated by the value to society, a company or the individual. Energy use that creates high value of any of these will in general be more difficult to replace than energy use that creates low value. It will be difficult to argue that a sales person or a manager should not spend some energy visiting an important customer to secure a large contract. This is an example of energy use that creates high value for society, the companies and the individuals involved. The use of energy to light, heat or cool a room that is not occupied or the use of oil to produce and distribute a piece of plastic packaging when none is necessary are examples of low-value energy use. All use situations could be valued along this type of continuum, even though the value of many types of energy use may be subjective or down to differences of opinion, such as the value of sending people into space or traveling to distant resorts for vacation. In both these cases, we also find that the activities create economic activity and they may spur economic growth, while people may still individually give them a low rating from the value that they bring to society or to individuals. This indicates that we may have to use multiple charts or that we will have to elaborate the model further to make it really useful.

Along the axis that measures the cost of reducing energy use we identify substantial differences between alternatives. The cost of reducing

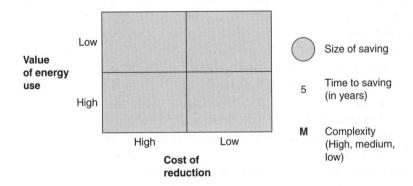

the energy use of lighting rooms that are not occupied is zero if the method is based on the idea that people switch off the lights when they leave. If we evaluate the idea of controlling lighting by automation, we find that the cost is slightly higher. In the model we illustrate the size of the resulting savings by the size of the cycle used to mark the position of the saving in the matrix. We would expect the saving to be larger in the automation case than in the case of manual control of lighting. In the case of reducing energy use by replacing vehicle fleets or production machinery the cost will be higher. The size of the cycle, as already mentioned, illustrates the size of the saving and the figure that accompanies the cycle illustrates the time from the start of a project until the bulk of the saving has been achieved.

The seemingly illogical order of the "high" and "low" alternatives is justified by the fact that, with this configuration, the most attractive savings opportunities will end up in the upper right square, which is the way that most four field matrices are configured. Managers have become used to finding the most attractive alternatives in this place and it makes sense to use this principle in this case as well.

In Part 3 different savings alternatives will be discussed and the matrix will be used in order to relate the different opportunities to each other.

9.3   GUIDELINES FOR STRATEGIC PRIORITIES AND RESOURCE ALLOCATION (Item 3 on the list of tools for Steering Committees)

In a large program, inevitably, some projects or subprojects are more important than others. There will be technologies that have the potential of delivering larger savings faster than others. More resources and tighter project management may be allocated to such technologies, than to less promising projects. In the case of unexpected delays in very promising projects, the steering committee may want to investigate options to rapidly change aspects of a project plan or the intended configuration of a product, so that the obstacle could be removed. Such rapid reorientation of projects can only be achieved if project plans, in addition to the plans themselves and breakdown into subprojects, also included a priority for each plan, and the alternative routes available to solve particularly knotty or time-critical problems.

The steering committee of a program, stream or project needs to make a risk assessment and think through the possible prioritizations that need to be made by managers. It also needs to define what the most important priorities are, in case they are not already defined by the overall values of the project.

CHAPTER 10

The Standard Tools of Program and Project Management

In order to achieve a tightly planned project, there is a need, in addition to the tools for the steering committee, for tight day-by-day and month-by-month management of the change process. This management starts at the high level with the management goals for society, set by the steering committee, and the application of a method for change management and tools to support this process. Such tools are available and they are used already in order to drive change in large companies and in public organizations. The current debate in society, which is based on the argument that a market approach to the energy transformation is preferable, overlooks the need for planned change efforts on the level of society. In order to assist in the preparation for an insight that will probably be inevitable, a number of tools are listed below, which describe their purpose and how they can used. Those readers who are already familiar with change management projects may have encountered some of the tools, or other tools that are similar to those described here. For readers, who are not familiar with the concept of change management, the level of development of these tools, and the wide usage of them in existing change projects, may be a novelty. If we are going to run the energy transformation program as a planned project, it will be necessary for many people in society to use project management tools in the near future and the list below may represent the starting point for the selection of tools.

10.1 CHANGE MANAGEMENT – A TRIED AND TESTED METHOD

The concept of change management has been developed by large corporations and consulting companies in the pursuit of improved operational effectiveness. It has also been applied to the development and implementation of new strategies, as in the strategic reorientation of GE, by former CEO Jack Welch. Welch took GE away from a number of mature businesses where the company had little opportunity to achieve high profitability and growth and into a number of high-growth areas with high profit potential. Among the businesses that GE moved into, or strengthened, under his leadership were high-performance plastics, financial services and information system solutions. The current CEO Jeffrey Immelt has, since 2001, continued his quest. He has divested businesses to a value of 50 billion dollars and acquired new ones to a value of 100 billion dollars. He has also, among other things, started the project Ecomagination, with the aim of developing products and services that will be needed in the future when oil is scarce and an increasing portion of humanity lack fresh water. The strategic reorientation of GE has not only been accomplished through divestments and investments, but it has also involved tremendous efforts in the field of change of organization, work practices and strategic decision making.

Box 10.1 GE Ecomagination – A High-Level Plan for Positioning GE in the Growth Segments of the Future

As one of the world's largest companies, GE needs to focus on large-scale business opportunities. Through its overarching value of being a global leader in a number of high-tech growth industries, this company has identified providing solutions to the resource problems of the future, as an opportunity for sustainable business growth, both from a business and an environmental perspective.

Therefore, the CEO of GE, Jeffrey Immelt, has created a new project, called Ecomagination. According to the analysis of GE, the ecological and resource problems in the world, among which global warming and supply of oil are two of the most prominent, but where the impending scarcity of water will become increasingly important during the coming decades, the underlying reason for global predicaments is population growth. This growth leads to problems of overconsumption and scarcity of key resources. Overconsumption in our energy-intensive society leads to increasing use of energy, which leads to CO_2 emissions which in turn lead to climate change. In order to develop the technologies and products that will be needed in the future, GE has started to invest heavily in the development of ecologically friendly alternatives to existing technologies, and new resource-efficient generations of existing technologies and products. Within this program, GE employees are encouraged to develop technologies that will provide future solutions for the production of energy, such as wind power, cogeneration of electricity and heat and a new engine generation for trains and airplanes. In this project, GE has also started the development of technologies that, in the future, will provide fresh water to areas that experience water scarcity. It is expected that, while currently 92 percent of the world's population has access to fresh water, this figure will be reduced to some 60 percent by 2030.

In total, GE started with an investment of 700 million dollars in Ecomagination in 2004, and the annual investments will be increased to 1.5 billion dollars by 2010. The products that have been developed in the project have, so far, met with a much higher demand than was originally expected, and the market demand is expected to increase even more in the future.

Change management has also been applied in order to improve operational effectiveness. Even though many critics argue that the success of efforts such as business process reengineering, lean management and supply chain management has been mixed, the change efforts that have been made would not have been nearly as successful as they have been, had companies not developed competence in the field of "change management" and the tools that are contained in this concept.

Below, the theory and practice of change management will be described and the contents of a change management program will be outlined.

10.2   GOALS AND KEY FIGURES (Item 1 in the list of tools for project managers and participants)

It has already been argued above that goals are critical in a change process, such as the transition to an energy-efficient future. It is crucial that the participants can focus on common goals and objectives. For people outside of the project, goals make the change process easy to understand and it becomes possible for managers of a process to communicate goals and the steps that need to be taken in order to reach those goals. The goal of producing 50,000 airplanes per year during the Second World War is an example of one high-level goal set by President Roosevelt before entry into the war by the United States. This was later adjusted and broken down into production goals for bombers, fighter planes and other types of planes, and concrete production targets per month or year were set for each plant that produced planes. In today's business, we have more experience from change management than managers and politicians had in the US in 1941. It is also the case that today's products require more specialized production facilities and tooling than those of the Second World War. This means that the rapid transformation of production to a completely new line of products for the war will be difficult to repeat for the energy transformation program. What we can learn from the war effort and the Apollo program is that through planning it is possible to achieve goals that cannot be reached by market forces alone. We need to adapt the details of each of these experiences to current circumstances.

In a change process that will continue for several years it is important both to have long-term goals and also to have goals for the shorter term. Goals may be broken down for different dimensions and time frames. In the case of energy efficiency an overall goal may be set for the reduction of energy consumption or CO_2 emissions in a country. This goal may in its turn be broken down to the reduction of the same parameters by region, by industry and by plant, in order to arrive at an operational plan at the plant level of which measures need to be taken in order to achieve the goal.

The goal may also be broken down into investment in, and production of, new power generators that use renewable energy sources and this number may be broken down into the number of photovoltaic cells, wind power generators and other renewable power plants and generators that need to be built. Goals may also include new technologies that need to be developed, such as more efficient technologies of the above-mentioned types, and, possibly, new hybrid engines for trucks and more energy-efficient components and systems for industry and new control technologies for the electricity grids of the future.

10.2.1   The Process of Setting Goals

As mentioned earlier, in a market economy, each company is allowed to pursue the strategies it wants and to set its business goals for itself. The point is that each company should thrive only if it succeeds in catering to a market need. If there are customers who are willing to pay for its products, then there is a need for them. In a market economy, each company makes plans for its business, based on forecasts for the future. These plans are made in the face of a certain amount of uncertainty, because it is never known how much money the customers in a market decide to allocate to purchasing products or services from a particular company. This is determined by thousands or millions of individual customers as they decide which products and services to purchase.

A market economy is based on the idea that resources are best allocated by "the invisible hand of the market." In a consumer market, this makes a lot of sense, as we have discussed already, because different consumers demand different products and services, and because tastes and preferences change more rapidly than any planner could possibly foresee. The needs and wants of consumers also determine which investments are to be made by companies that produce consumer goods and also by the companies that produce the machines, which make consumer goods, and so on. Machine manufacturers making machines that are not efficient enough to produce at a competitive cost or that are not flexible enough to allow for frequent enough changes in a product will soon go out of business. This process, by which less efficient companies are replaced by competitors that are better at satisfying consumer needs, is sometimes called "creative destruction," and it was first described as something positive and necessary for economic development by the economist Joseph Schumpeter in the 1930s.

As in the case of World War II administration, this turned the American economy into a planned economy, almost overnight. The military and its soldiers were not in a position to be picky about which type of rifle they wanted or the make or model of an aircraft. The most important aspect of production, in this case, was to supply the armed forces with the best products possible, produced in the largest possible volumes. American fighter pilots could not argue that they would rather have a Spitfire, made in the UK, and preferably a blue one. They had to settle for the models that were available in the battalions to which they were assigned. Compared to the situation in a market economy, the planned economy of war materials production provided a predictable demand for the War Production Board and for the companies that produced the material. This was just as well, because all producing companies had to work at maximum speed, turning out the largest number of aircraft and other materials possible. There was no time to wait for the users to react to and evaluate product features.

In the same way an overall plan of the US space program, administered by NASA, the space program equivalent of the WPB, which reached out to suppliers of technology, products and services across the American economy, was the foundation of success in this case. How could NASA have succeeded to put a man on the Moon, had it not been for the overall plan, which included a picture of all the key systems and subsystems that had to be put in place, by 1969? How could NASA rely on the market to produce a propulsion system for the Apollo program, or any other system, without attempting to manage the process of developing this? In order to achieve a rapid development schedule, program managers had to decide at an early stage of the project, which technologies to use and how different technologies would interface with each other.

Clearly, neither the planning during the war, or the planning of the Apollo program, nor the strategic and operational planning that on a daily basis is going on inside companies, have made either the US or the European economies planned economies! Instead, these instances of planning have made it possible to achieve impressive feats of economic development in the history of civilization. Planning facilitates the achievement of the "big hairy audacious goals (BHAGs)" that James Collins and Jerry Porras, in *Built to Last*, identify as one of the key prerequisites for sustainable success in business.

The transformation to sustainable energy systems could be seen as the next BHAG for a global society that is used to achieving amazing industrial and economic feats.

10.3 DEVELOPING A STRATEGY FOR THE ENERGY TRANSFORMATION (Item 2 in the list of tools for project managers and participants)

The development of an overall strategy for change is a very important aspect of the transformation program. In order to develop such a strategy, all the relevant knowledge, experimental data and analyses of existing and future sustainable energy technologies need to be gathered and analyzed in order to determine which paths forward that are most promising in each stream of the program.

One other important aspect of strategy development for this purpose is to determine which areas need to be, or should be, managed in the planned change process and which areas should be left to the market forces to take care of. A strategy should also determine the financial solutions that need to be used in order to make available the financial resources for the transformation for each aspect of the change. In which cases does government financing represent a necessary alternative, and which parts of the program can be handled by the financial markets and the companies left on their own? Which technologies used by companies and which consumer products need to be subsidized? Is there a need to levy taxes on some existing technologies or, particularly, energy-consuming products or services, in order to make energy-efficient alternatives more competitive?

All these aspects and more need to be covered by a strategy for the energy transition of each country. It is possible that a lower degree of planning could be tried and implemented at first, covering only a small number of prioritized areas. The plan could then allow for further areas to be planned or regulated, if transition pressure increases or if results are less promising than expected. This would be similar to the implementation of rationing programs during a war. At any one moment, only products that had to be rationed were rationed, but preparations were made to implement new rationing programs as the need arose.

We need to reason, in a similar way, regarding energy transition. Planning and other government intervention should be confined to areas that need rapid progress. Measures should not be stronger than necessary. In some or all cases guidelines, information and training may suffice, but there may also exist a need for stronger intervention. We cannot possibly know the exact need for these measures until we have made a strategy.

10.3.1   Developing a Strategy for New Technologies, Products and Services

One strategy is a high-level road map, showing the overall direction of development and for the different streams of the change program. The development of a realistic strategy for the energy transformation will require substantial amounts of analysis at many levels of the overall change program, and in companies that will contribute to the change efforts. This strategy development will require in-depth knowledge of a number of different aspects of the change program. We need to understand the implications for the short and long terms of different alternative technologies, and the volume of energy that different alternatives could contribute on different time scales. In Part III of this book, different technical options will be evaluated from a change management perspective. This has not been attempted before, and this book will not be able to provide any final answers. We need to start to ask the relevant questions, and urge people who are in the position to make decisions about this, to go further down this path of enquiry.

10.4   A PLAN FOR CHANGE (Item 3 in the list of tools for project managers and participants)

Based on the description of the US effort to win the Second World War and other experiences of large-scale change, a framework for change, which could form the backbone of a strategy to transform the current energy-consuming systems of the global economy, is presented below. The task in itself is a truly massive undertaking, and leadership will be needed at different levels of the program and in all sectors of the economy and in society as a whole.

The overall plan, and the lower-level plans that need to follow from it, will be key to the transformation effort that is to come, and all of them will be crucial to the success of the endeavor. One aspect that planners, at an early stage, will have to take into consideration is what degree of planning will be relied upon during the change, and which financial mechanisms will be used by the government to support the development toward sustainability. Later development could increase the degree of planning and put new mechanisms to work, which can increase the precision or the speed of the process.

Some degree of planning of the process will become necessary. This could mean that a central authority or nongovernmental organization takes on the responsibility of building knowledge and competence related to change management issues in the process. This would be a low degree of government intervention, or planning. A higher degree of government intervention, or planning, could involve price controls for renewable fuels, or the promise by the government to order certain numbers of products that contain a new technology, or the direct investment in development efforts or other measures to rapidly increase supply of renewable fuels or sustainable engine technologies. Based on the urgency of the situation it is reasonable to believe that a high degree of planning will ultimately be needed. One problem in achieving this will be in gathering the political will to change as well as the courage to finance such a change. It is unlikely that all this will be achieved in one political step. Rather, it is more than likely that a stepwise approach will be necessary.

This means that we not only need to build competence in technical issues, as has largely been the case so far. We now have experts who understand many of the issues related to ethanol production and the production of other renewable fuels. What we lack is an idea of what the future mix of alternative fuels will look like and how we are going to make sure that the supply of each of these fuels will match future demand. In addition to this, we need an understanding of which engine technologies will be necessary, and how many vehicles we need in order to use the renewable fuels that we will be able to produce. We also need a rough idea (if we decide for a high degree of planning we need a more detailed idea) of which steps are needed in the development and which companies are going to take different steps and at which stage in time we could expect a certain step to be taken.

We could use the debate around ethanol as an example. At one stage ethanol was described as a very promising alternative to gasoline. In the short run, however, ethanol is produced from grain and sugarcane, and grain-based ethanol requires that much oil be spent in agriculture, producing the grain. Therefore, grain-based ethanol is not a very energy-efficient alternative in the long run. A better alternative would be cellulose-based ethanol, but we lack an economically viable process for this production. There are small-scale plants, which however have a relatively high production cost. In the future, though, we will have to produce biofuels of some kind from cellulose, since we will not be able to use enough of our arable land to produce energy crops. There are, however, a number of different cellulose-based fuels available, which we need to analyze in order to select the best alternatives. Methanol is one of them, and dimethyl ether (DME) can also be produced from this raw material, but we lack large-scale production facilities for both of these and for DME we lack a fleet of vehicles with engines that can use this fuel. Methanol could be used in a mixture with gasoline, with up to 30 percent methanol, without modifications of the engines. These alternatives will be discussed in more detail later. At this point, we only need to recognize that a plan will be necessary in order for people and companies to be able to make decisions about investments and choices.

In order to achieve this, we need to build demand. This is at present done by producing and selling ethanol-fueled cars, trucks and other vehicles. We need to run those vehicles on the ethanol that we have for the moment in order to build demand for other sources of ethanol and other renewable fuels. At the same time, we need to develop an idea of how much ethanol, methanol or other fuels we can expect to produce from cellulose in the future. We can do this by analyzing the amount of forest that we have and the growth rate of this forest, and when large-scale production facilities for cellulose-based fuels can come online. Furthermore, we need to determine how much we can use for fuel and which countries will demand this fuel in the future. There are still large forests in the Northern Hemisphere, in Canada, Scandinavia and Siberia, and substantial amounts of fuel could be produced both from wood and from residues of forestry. If a large part of all cars are going to be run on cellulose-based fuels in the future, how far will this supply last and how much of the forest are we willing to use for transportation? Estimates indicate that substantial volumes of cellulose-based fuels can be produced in the future, but that the production will require hundreds of large-scale plants, which will have to be scattered across each continent for this scenario to materialize. Each of these plants will have to be supplied by some 450 truckloads of wood every day, which will have to be transported from the wooded areas to production facilities close to the markets.[1] Alternatively, the plants need to be located close to the sources of raw materials and fuels need to be transported to the markets.

One other issue is regarding how the ethanol is going to be distributed. In some countries, such as the United States and Sweden, gasoline is already mixed with a certain amount of ethanol. This amount could be increased to the level of some 30% without changes to the engines. Mixing biofuels with petroleum represents a simple and inexpensive way of distributing ethanol and this increases the environmental friendliness of transportation without creating a separate and new distribution system for ethanol, requiring separate pumps at gas stations, separate deliveries of ethanol and gasoline and other sources of increased cost.

The downside of the mixing solution is that no real competition is created in the fuel market. The ethanol producers simply become suppliers to petroleum companies and in case we would like to create a future fuel market where two, three, four or more alternative biofuels compete with gasoline, we should go for a solution in which each fuel is sold separately. This would, however, be achieved at a higher cost, and, depending on the financing of the project, possibly a tremendous risk for the companies that will develop the fuels and invest in the distribution systems for them. Once more, the analysis and strategy development will need to determine how much time we have at our disposal and which alternatives will it be possible for us to choose from. A certain share based on blending solutions will be a relatively quick fix and separate distribution systems for each fuel represents the more ambitious and capital-intensive alternative.

In order to decide on the choice of solution, which in each country to a large extent could be made into a regulatory issue and a planning issue, we need to have a vision of the role the different alternative fuels will play in the future fuel market and which alternatives we foresee to be present in 10, 20, 30 and 50 years. After all, the most viable and economic future solution for automotive fuels may, for all we know, be electricity, and recharging may to some extent be done through photovoltaic cells. The fuel of the future may also, as some experts argue, be hydrogen (which will be used in fuel cells), which is produced from water through electrolysis, a process which in itself requires large amounts of electricity. Some advocates of electric vehicles argue that it will be much more energy-efficient to run vehicles on electricity directly by charging an electric battery on the vehicle, since three times the amount of electricity has to be used for electrolysis, compared to direct use of electricity for propulsion. However, at the moment, without a sound overall analysis, we can only guess at the outcome of this development. Most probably we will have to use a number of different solutions, but the possible future volumes and the time frame for the development and implementation of each of these still remain to be determined.

We already waste valuable time and money by debating different alternatives, without the ambition of developing and refining a plan. First, we need to agree upon the criteria that we need to use in the evaluation of alternatives. Secondly, we need to assemble the facts that we already have access to, but which are scattered around society. The next logical step will be to complement the data with data that we lack in order to finish the analysis. After this, we can finish the plan and get to work. In fact, as in the Apollo program and other large projects, we can get to work on the early steps even before the entire plan is finished. In Part III, I have started work on the plan, using data that are available. With more resources, I would have been able to gather more data, and provide a more complete analysis. The shortcomings of the attempt in this book are only indicative of the need for more work.

10.5 A PROGRAM AND PROJECT ORGANIZATION (Item 4 in the list of tools for project managers and participants)

Any change program, in a company or in society, needs a formal organization. In the case of the Second World War industrial transformation, the core of this formal organization was represented by the War Production Board and in the case of the space program, the formal organization is represented by NASA. Representatives from these core organizations, will, however, work in committees and project teams, together with representatives from companies that run projects in the programs, experts and other people who are involved in the program. These committees and joint teams and the people in companies who work in the projects are also part of the formal organization of the program. As mentioned above, it has been estimated that 400,000 people were involved in the Apollo program. The work of each of them needs to be organized.

In business, the organization of change projects has changed remarkably during the past two decades. Twenty years ago it was normal for large companies to appoint strategy consultants for analysis projects. These companies, typically, delivered a report with a thorough analysis, telling company management which of the strategic opportunities it should pursue and containing perhaps a plan with the steps that could be taken in order to fulfill the strategy. The steps could consist of focusing on a certain business area, divesting businesses in other areas, acquiring certain companies in order to strengthen the position in the core focus areas and a number of "dos and don'ts" for the change process. The company would then have to manage the change by itself, and, needless to say, in many cases very little happened after that. The analysis and planning effort may have involved a team of five consultants, some part-time on the project, with different tasks and specialties, and the duration of the project would have been 2 to 6 months, depending on the complexity of the issues and the demand for detail. Employees in the company that appointed the consultants may have been involved in interviews with the consultants, and high-level managers may have been invited to presentations of the results.

Modern change projects in business often involve many more people and the people involved spend more time in the projects. It was noticed that in the type of projects described above, the speed of change was reduced, the motivation of employees waned and the knowledge of what to do and how to do it disappeared when the consultants left, which often happened after the analysis was finished. Employees in the company organizations often did not have the skills or the drive that would have enabled them to run the whole project on their own. They were often so bogged down in their daily tasks and they lacked the knowledge and the tools to do completely different things, such as evaluating acquisition targets, and acquiring companies, or preparing existing parts of the business for divestment. Some organizations had these skills, but many did not. As we will see below, it is truly difficult for any person to learn new skills and to get to work with new tasks, or using new tools.

Change management requires substantially more resources in the change program and in most companies it also requires more involvement from consultants and for greater lengths of time, than traditional strategy consulting. Companies that run change projects repeatedly, such as GE or Nike, have employees who have developed skills and have gained experience that is necessary in order for employees to be able to manage or participate in a large change program. Many organizations that have less experience from change or run such projects infrequently have not.

Realizing that traditional strategy analysis often does not lead very far in terms of actual change, change programs today are often organized so that the same core team of persons will be able to participate in a project from start to finish, so that some of the participants in the analysis phase also become the leaders in the change project or program. The analysis, or design, phase is typically performed by a small number of people, who set goals for the change and make a plan for the large change program. In the change program in a large organization, there will be a number of different streams, with one manager from the company that functions as the person responsible for each stream. There may be 10 or 20 streams that may cover such things as IT systems transformation, modularization of product design, change management in different departments in the organization and communication streams, which work with the communication of plans and results to employees outside of the project, and to remote offices.

In change programs in organizations, which are relatively inexperienced in change, each manager for each stream in the program may have a full-time or part-time consultant supporting her/him with the management of the program, preparing materials for decisions, preparing the manager for meetings and other important skills that the manager needs to acquire for her/his role. In some cases this support may last for the entire program, which may take 12 or 24 months to complete, or the consultant support may be reduced as the manager learns to manage the program or stream without support. In addition to this management structure, there will be employees without a management function in the program, who administrate, perform analyses or have other roles in the project. There may also be consultants with specialist skills in particular areas.

All in all, this type of project may require the involvement of 50, 100, or several hundred persons from the organization of the company that runs the program, with support from sometimes almost as many consultants from one or several external consulting companies.

The advantage of running change programs in this way is that the organization is dedicated to running the program and many times the people involved are with the project full-time. These managers and employees will be some of the most skilled and motivated people in the company and they will have the support from even more specialized consultants. This ensures that the project is driven with much energy and that the effort is goal oriented.

The cost of running this type of project may be several million dollars per year. For a large company that is in need of strategic reorientation or some other major change effort, this cost can be justified by the rapid adjustment to a new strategic focus. A focused change effort does not ensure that the goal will be reached. For several reasons, this type of program may fail, but the structured approach of change management highly increases the odds of success. For many companies in highly competitive industries that are badly in need of becoming more efficient, rapid change may be a matter of survival. In this type of situation it is often better to spend a large sum of money to rapidly come back on track, than to take the risk of running the change program with insufficient resources and run the risk of failure.

In the energy transition, the strategy analysis may identify the need both for a strong organization on the part of society as a whole and for strong change management efforts in many companies. As indicated, this would be a very large investment for organizations that need to or want to transform themselves to sustainability. It may be that this type of change will not need a large full-time organization in most companies, but it may be that some companies will need to change in a rapid and focused way. This may both be companies that need to remodel their products or services in order for them to become more energy-efficient in order to generate more demand from the market, and there may be companies that need to improve their own business processes so that they, themselves, become more energy-efficient, or both. Hopefully, most companies will not need to undertake a massive effort and it is not the goal of this book to argue for change or large change programs for their own sake. We need to expect, however, that large-scale managed change will be one of the options for the future.

10.6 MANAGEMENT SYSTEMS, PRINCIPLES AND GUIDELINES (Item 5 in the list of tools for project managers and participants)

In addition to the formal organization structure of the change program, the process needs a number of firm management mechanisms that need to be put in place in order to run the change process.

The program organizations need internal meeting structures, pre-established agendas for meetings, checklists for decision points and checklists for the preparation of decision material and guidelines for managers on how to prioritize in different situations, for instance between speed and attention to detail, and between energy transformation and financial goals. Such guidelines need to be based on the overall values that guide the change process, which can be determined during the strategy development phase, described above.

Examples of these tools exist in different forms as general tools that have been used in previous change projects in different environments, but they need to be adapted for this purpose. Adaptation, as already indicated, needs to be done based on the overall strategies and goals of the change program itself.

At this point, the lengthy discussion of the change tools is postponed (item number 6 in the list of tools for project managers and participants) for the moment and we turn to the final few tools for project managers and participants, which will be covered briefly. We will then turn to the change tools in the next chapter.

10.7 A COMMUNICATION STRATEGY (Item 7 in the list of tools for project managers and participants)

It is always important to keep the participants in a project well informed. This is not only true during a war, when information is particularly important in order to avoid disinformation and the spreading of rumour. There is a similar risk in large change projects. If there is a lack of information about progress or the outcome of important events, meetings or decision points in a project, this will be interpreted variously by participants. The effects of such interpretations need to be managed by program managers.

Thus, it is important to plan information activities in advance. Such a plan must contain directives regarding when and to whom particular pieces of information should be provided and how information to different interest groups and participants is going to be distributed.

In addition to this, a communication strategy could contain directives about who should be the sender of a particular type of information to a certain group of recipients, and particular information about the handling of classified or sensitive information.

Box 10.2 7×7 – A Rule of Thumb

Many companies that have run change programs have experienced that managers in the project tend to underestimate the need for information and communication with participants in a program. Even though people have been asked to attend one information meeting, it turns out that they still are unaware of many important aspects of the change program. This lack of awareness sometimes results in an inability to make decisions, or it may lead to wrong decisions and mistakes, which could cost a lot of money to correct.

In this respect, GE applies the rule of thumb that any person, who is going to participate in a change program or project, needs to receive information seven times each, from seven different persons in the organization. This way, a person not only receives information, say, from the project manager. She also receives information from the marketing manager, regarding how the change affects the marketing department, from the purchasing manager, regarding the impact of the change on purchasing and logistics, and from the R&D manager, regarding the aspects related to R&D. From each manager, each person not only receives information, but they also get the opportunity to ask questions when they have thought the matters through, and they get the opportunity to have another discussion with these different persons, when the project has been in progress for some time. All in all, the point is not that each person needs to be informed exactly 49 times, in the pattern described above. The important point is that each person needs to be informed several times, and that they also will need to discuss the program with a number of different persons, based on their experiences as they become involved in the work. In a smaller organization, the number of people to discuss with will be fewer than seven.

10.8 TRAINING AND PREPARATION OF TRAINING MATERIALS (Item 8 in the list of tools for project managers and participants)

In order to transform energy systems, the participants of the program need to be trained, and in other ways prepared for their tasks. This training will require both time and resources and it could be seen as a bottleneck for the initiation of the program. A large part of the training effort can only take place once the strategy, organization, management principles and the tools have been developed, or selected. When these basics of the program have been put in place, training materials, literature, presentations and work material need to be developed.

In order to facilitate a rapid scale-up of the program, training can be managed in a number of "waves." The first wave would consist of "training the trainers." This could mean that some of the people who have participated in the project planning up until then, and a number of professional trainers, take on the task of training a number of experienced consultants and change managers, who in their turn are assigned the responsibility of training other key people for a large-scale effort.

The training of the main part of project participants could take place as the first steps when the program organization has been put in place, so that people who are going to work together also could be trained at the same time.

10.9 DETAILED OPERATIONAL PLANS FOR THE IMPLEMENTATION (Item 9 in the list of tools for project managers and participants)

If all the preparations have been done in a conscientious manner, the implementation of the program and getting all the tools and routines in use will become easier, but project implementation will still be the most difficult part. To change the work practices of a large company takes a lot of time and it requires tremendous monitoring of progress and attention to detail. The initiation of a large-scale program at the level of society will be even more difficult.

Implementation of new routines is a matter of attention to detail and repetition. The work practices of individuals and work groups need to be monitored and corrected, when they deviate from the management principles. Managers also need to set aside time for the necessary management and information activities of the program. An important role of management, in this phase, is that of relentlessly communicating the values of the change program, how participants should prioritize and what the overall expectations for the program as a whole and for each particular stream or subproject are. If the tools and practices are closely linked to the values and the goals of the project, communication and management will be a lot easier.

One important task of the managers of streams in the program, will, at this point, be to break the overall plans of the program down to more detailed strategic and operational plans for the particular projects that each stream will contain. Each project manager then needs to break her project down into subprojects or project phases and make detailed plans for those.

CHAPTER 11

Change Tools

11.1 ITEM 6 IN THE LIST OF TOOLS FOR PROJECT MANAGERS AND PARTICIPANTS

Many things need to be included in the toolbox that has to be available to the participants of the change program. The need for tools ranges from overall principles for how a strategy for change at different levels is to be developed, over a theoretical framework for how individuals and organizations learn to adapt to new situations, to more specific tools that can be used in this adaptation process. In this chapter, we will discuss a number of tools and also discuss how they can be used in order to facilitate change in the energy transition program.

Some of these tools have been developed over the course of writing this book or adapted from other existing tools.

The first tool is a theoretical model of organizational learning, which, compared to other existing models, clarifies the two roles of the individual and the organization in a process of organizational learning.

11.2 A MODEL OF ORGANIZATIONAL LEARNING FOR ENERGY TRANSFORMATION

11.2.1 Organizational Learning – A Background

A change process could be seen as a large-scale training effort, through which a large number of people learn to perform tasks in new ways by changing their values and their behavior in a planned and structured way. The overall planning of this change is often done by management, but it is influenced by people from all over an organization and from the organization's environment. Change objectives are often communicated to the organization in the form of overall goals that are gradually broken down into subgoals per unit or by employee.

There is a vast literature on organizational learning. The most prolific researcher in this field, worldwide, is Peter Senge, the author of *The Fifth Discipline*. Senge's work has been inspired by theories from the 1970s that were developed by Chris Argyris and other researchers.

Argyris made the division between single-loop and double-loop learning. Single-loop learning takes place within an existing frame of reference, while double-loop learning is the process of developing a new frame of reference, within which a new series of single-loop learning cycles can take place.

Senge developed this theory into a new framework for individual learning and the learning of teams. He identified five "disciplines" that are necessary for individuals to master, in order to be able to learn the new skills that they need in the workplace of the future. These disciplines are

- personal mastery,
- mental models,
- shared vision,
- team learning, and
- systems thinking.

Senge argues that only the individuals in an organization have the means to learn anything. The basic unit for learning in the organization, Senge argues, is the team. Teams learn new skills collectively, because the individuals in a team learn how to perform their own tasks and they also learn how to interact with their peers.

In *The Fifth Discipline*, Senge does not explain how team learning and individual learning are related to organizational learning. Where is the organizational knowledge "recorded," and how is the "recorded" knowledge retrieved by individuals, who need to access it in order to perform their daily routines? The documented routines and procedures of an organization, which capture a small percentage of the total knowledge of an organization, may seem far too prosaic to encapsulate the collective wisdom of an entire organization. Where, then, does the knowledge of the organization reside, and which forms of knowledge are present in an organization?

I believe that it is important to answer this question in this book. This is because the amount of learning in society and organizations that will be required during energy systems transformation is very large and the speed with which it needs to take place is very high. The process will require a robust and down-to-earth understanding of how the process of learning in society and organizations works, and this has to include an understanding of the relationship between individual learners and the organization as a whole.

In my view, organizational learning also involves changes in the structures of organizations and society. The learning of individuals of how to perform tasks is different from the learning of organizations, which is captured by the different routines, documents, processes and organization structures of an organization.

11.3 A NEW MODEL OF ORGANIZATIONAL LEARNING – MERGING THE INDIVIDUAL AND THE ORGANIZATIONAL PERSPECTIVES

The units in an organization that have the ability to learn anything are the individuals and the organizations themselves. The knowledge of individuals does, however, consist of more than mere detailed knowledge of markets, customers and products (which constitute examples of the material knowledge of employees in marketing) or knowledge of raw materials, components and manufacturing technologies (which represent examples of the material knowledge of employees in production). Knowledge in an organization also consists of "operational knowledge," which includes knowledge about how different individuals, teams and departments work together, how tasks could be divided between individuals or teams or how a certain production routine is best performed. Individual knowledge also consists of "strategic knowledge." This includes knowledge and an understanding about which strategy alternatives work for a particular company, because of such things as its competence base or cost structure, and which new business opportunities may be pursued by a particular company.

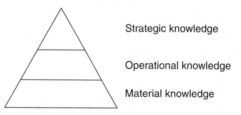

Strategic knowledge

Operational knowledge

Material knowledge

Figure 11.1   Three levels of individual knowledge

Each individual learns new things through interaction with other people, or by taking in information from books, newspapers, information leaflets or other written material. We also learn new things by structuring our knowledge of a subject in our own minds, in the way that I did when I wrote this book. The most obvious source of learning in a tightly knit organization, working day by day together, is the team. Teams discuss tasks and the roles and contribution of different individuals to the performance of the team. In any organization, where teams and individuals have relatively fixed tasks, individuals will primarily learn from other team members.

The learning of organizations is captured in the form of the structural capital of organizations. Some of the material knowledge of individuals will be captured in business systems, work descriptions, contracts and other documents that are owned and used by the organization. Some of the strategic knowledge will be captured in strategic plans, business plans and other documents, and some of the operational knowledge will be captured in process and work descriptions and other documents that belong to the organization. Regardless of the amount of effort that is put into the explicit and formalized documentation of "organizational learning" there will always remain large amounts of knowledge that rest with each individual in the organization. This knowledge, or skills, may exist in the form of tacit knowledge about things that have not yet been documented or knowledge about documented principles, routines or procedures, which the individual has "internalized" and, sometimes, virtually forgotten about, because they have become an integral part of the individual and her way of functioning in the organization. The skill of driving is an everyday example of something we learn, which we have internalized and "forgotten" and converted into tacit knowledge. The ability to learn skills, make them tacit and forget about them seems to be an important factor in the learning process, because once we have internalized a piece of knowledge, we become open to going through a new cycle of individual learning.

The model of organizational learning below integrates individual learning from tacit to explicit knowledge and into the realm of organizational knowledge, which could be incorporated into the structural capital of the organization. The structural capital is then used more or less skillfully by different individuals. The way that the structural capital is actually used determines the organization's frame of reference.

It is then the experimentation and learning done by individuals, first represented by tacit individual knowledge, which forms the basis for the next cycle of development of an organization's or society's formal knowledge and structural capital.

Based on the above reasoning, the following model of organizational learning is proposed:

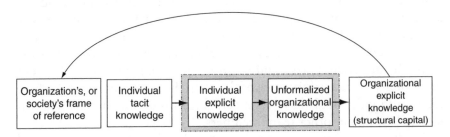

The model in itself can be applied to any situation in which we can observe a combination between individual and collective learning and knowledge development. I will apply it to the learning that will take place inside organizations and that will be required in order to develop energy-efficient products, services, operational processes, procedures and corporate strategies.

11.4 THE DEVELOPMENT OF STRUCTURAL CAPITAL IN ORGANIZATIONS AND SOCIETY

11.4.1 Structural Capital in Organizations

The development of structural capital in organizations is, according to my point of view, an amalgam of a number of different theoretical subjects, starting with organization theory, covering strategic planning, IT and operational efficiency. In a classic passage in *The Wealth of Nations* (1776), economist Adam Smith describes how the division of labor in a pin factory dramatically contributed to efficiency in pin manufacturing. This division of labor represents an early example of structural capital. By structuring the process and organizing work in a number of steps, the entrepreneur in this company created a structure that provided the company with a competitive advantage over companies that employed a less elaborate process.

Around the turn of the twentieth century, the American Frederick Taylor introduced the concept of "scientific management." He argued that production could be made much more efficient, compared to the state of the art at the time, if tasks were analyzed and broken down into subtasks and divided between different employees in a scientific manner. At this time, the formalization of tasks was not by necessity made into written instructions used across the organization. They were many times communicated through word of mouth between experienced workers and their young apprentices. Nevertheless, the division of labor that was developed represented structural capital.

Throughout industrialization, machinery to a large and increasing extent helped in the development of structural capital. Initially, machinery was based on standardized technologies that could be used for many purposes (such as lathes, typewriters and conveyor belts) and products and production processes could be configured in very different ways, using the same machinery. (This is probably the primary reason why it was possible in 1941 to so very rapidly transform the US economy into a war economy.) Two different companies could then, for example, develop cars that were constructed in very different ways. While the general purpose of the two cars was the same and both used the combustion engine for propulsion, the design, manufacturability and the serviceability could be very different. Each producer and product had a few competitive advantages and also a number of weaknesses, which owners gradually became aware of. In the 1990s companies started to increasingly learn from each other and "best practices" were developed. Best practices could be described as a form of structural capital that is available at the level of industries or in society as a whole.

Products, manufacturing technologies and construction solutions, as well as organization structures, represented the structural capital of an organization. For a great part of the twentieth century, there were few formalized and widely used methods for the development of new products, organization of production, structuring of a business organization or for the development of strategies.

As organizations grew larger, a need to structure them was developed and the work of Alfred Chandler and Alfred Sloan at General Motors and similar efforts in other companies resulted in the development of the divisionalized organization. The need for large companies to specialize in areas where they had the best competitive advantage resulted in the need to develop strategies and tools for strategy analysis. Thus, organization structures, strategies and strategy development processes were incorporated into the structural capital of organizations. Gradually, some companies, such as GE under Jack Welch, developed strong formalized processes for strategic reorientation, which have become part of the competitive advantages, and structural capital, of these organizations. Nike and others have developed strengths in entering new markets, such as swimwear, football shoes and golf equipment.[1]

During this development many forms of work descriptions and instructions have been developed to support employees in their daily attempts to replicate successful practices of other people in their own organization or replicate best practices from other organizations. It is a key competitive advantage of successful companies to not only document successful practices but also to train employees to use them. With the development of quality management systems, such as ISO systems, including corporate quality manuals and other instructions that cover procedure both on the strategic and operational level, both the quality and consistency of structural capital and the ability to get it used in organizations are strengthened.

11.4.2   Structural Capital in the Public Domain

As companies and organizations have advanced in their efforts to develop methods, the processes and tools for the development of structural capital have been made increasingly elaborate. The leading organizations in each field, such as GE and Nike, have developed methods and strategies that create and sustain competitive advantage. The abilities of these organizations to identify new areas of opportunity, enter into these areas and build a profitable business in new business areas represent examples of structural capital that it takes time for competitors to copy.

Over time, however, the competitive advantages of the leading companies "leak" into the public domain and it becomes knowledge that increasingly becomes accessible to all companies. The same is true for features of products and services, business processes and other structural advantages. Through employees that change employers, consultants who transfer knowledge and best practices between non-competing customers, benchmarking efforts and research, business and technical literature, successful practices are transferred between companies.

Through university courses, MBAs and other training efforts, what previously used to be corporate structural capital of the most advanced companies is transferred into the public domain. In the case of increasingly specialized and turnkey production equipment, programmed industrial robots, modularized components that are offered in open markets and other equipment, the knowledge about industrial processes, product concepts, strategy methods and other competitive advantages of companies are transferred to the public domain. Gradually, the body of structural capital that is publically available grows.

11.5 THE APPLICATION OF THE MODEL OF ORGANIZATIONAL LEARNING ON ENERGY TRANSFORMATION

This model of organizational learning depicts an important process that many people who take part in the energy transformation will have to learn how to speed up. For a number of different reasons it will be important to understand how the process of organizational learning works in detail, and how individual knowledge gradually becomes organizational knowledge, structural capital, and structural capital that is available in the public domain:

- The energy transformation program in society will require a massive amount of organizational learning, both at the level of companies and at the level of society.
- This organizational learning will have to be based on knowledge that is currently being developed by a small number of individuals, who try out new technologies, lifestyles and work practices in their daily lives – at home or at work.

- These individual experiences will have to be collected and made into structural knowledge on a large scale by developing new equipment, infrastructures and services that support individuals throughout society in reducing their energy consumption.
- At the same time investments in large-scale energy technologies will have to be made in order to produce increasing amounts of energy from renewable sources in the future.
- To sum up, the learning processes that have to be initiated will have to bring forward new structural capital in different forms:
  - New values among individuals and businesses
  - Technologies, products and services
  - Work practices and processes
  - Changes in organizations and industry structures
  - Changes in society's systems and infrastructure
  - New and improved business strategies

Through the rest of this book, I will describe the nature of change in different areas, and the volume of analysis and learning, that will need to take place in order to succeed with the energy transformation.

Up until now the emphasis in companies has been on improving operations with the aim of reducing time and cost in each process. This has led to a situation in which we now realize that our lifestyles consume too much energy. One important reason for this is the fact that we, in our search for inexpensive production resources, have moved production further away from the marketplaces of Europe and the United States.

As we now need to reduce energy consumption, we find that very little knowledge about how to do this has so far been captured in the formalized business documents and products that make up the structural capital of most organizations. Even though energy-efficient processes will be one of the key competitive advantages for companies in the future, very little time and cost has, so far, been invested in capturing and formalizing this knowledge. Despite this, we expect that many individuals within organizations hold some knowledge about this as tacit individual knowledge. This knowledge could be made explicit and turned into individual explicit knowledge and also into structural capital for the organization. This could be done as a two-step process, in which the first step is to extract the individual knowledge that exists in the organization and through this identify key areas of the business to further investigate for energy improvements. The second step is to analyze the ideas with the best potential generated by employees, in order to prioritize the ones that should be implemented first and the ones that should follow.

During this process, key figures for the improvement goals need to be defined and improvement activities need to be devised with the aim of achieving these goals. In this way, the improvement of energy efficiency could be started and opportunities to improve energy efficiency at low cost could be identified. This is probably the first step in a long-term project toward energy efficiency.

CHAPTER 12

New Tools for Learning and Analysis

The traditional method for problem solving in our society involves an issue, an expert and a computer used for calculation and writing. The expert performs interviews with other people, who are experts in specific areas of relevance to the project. She also reads reports, books, searches the Internet and taps a number of other sources for information, which she analyzes, in order to draw conclusions about the subject matter that is being analyzed. These conclusions, and the most important points of the analysis, are written down and presented in a report, which is then distributed.

This approach is based on the way that scholars at the university approach a doctoral thesis or a research project. It is taken for granted that the best analyses are made by the most competent experts, a proposition that seems reasonable at a first glance.

In the case of developing forecasts for the future, strategies, or the selection between different approaches, the same method is most widely used. The "expert" may, in this case, consist of a team of consultants, a research team at a university or a team of experts at a company. This type of task is usually approached in a way similar to the one above. The experts make a list of interview questions, items to analyze, perform a number of interviews and analyses and analyze the result.

Increasingly, however, new methods are developed and put into use. Some of these are based on collective and parallel forms of problem solving, rather than traditional sequential analysis. In his book *The Wisdom of Crowds*, James Surowiecky argues that groups of people who solve difficult problems collectively often arrive at better solutions, than individuals, or traditional analysis teams. The issue, in this case, is that traditional approaches to teamwork often are biased because one team member or a part of the team takes the lead in discussions, and the team is led to agree on a mutually acceptable solution that has often originally been advocated by one, or a small number, of its members. In the type of collective approaches advocated by Surowiecky team members need to be unaware, or, at least, largely uninfluenced, by the reasoning, or the conclusions, of the other members, and multiple solutions and lines of reasoning must be allowed.

In his book Surowiecky relates a number of examples of this type of approach. One case is represented by the search for the missing US submarine *Scorpio*, off the eastern coast of the United States in 1968. On its way back to the base at Newport News, the submarine lost contact with the base. The area to search was a 20-mile-wide cycle, which was thousands of feet deep.

The leader of the search, John Craven, discarded the idea of appointing a team of experts on submarines and sea currents, in order to identify the most probable place to search. Instead, he developed a number of scenarios that represented alternative explanations of the disaster. Then he appointed a number of different experts in as varying fields as mathematics, submarines and salvage. He then asked all experts to offer their best estimates of the probability of each of the scenarios. He also offered them the incentive of winning a bottle of whisky for the best guess at each question. The experts bet on issues such as the reason for the accident, the speed of the submarine as it descended toward the bottom of the sea and the steepness of the descent. Craven did not believe that any particular estimate would offer the clue to where the submarine was located. Instead, he expected that if he put all the answers together, he would be able to come up with a good idea of where to look for it. The spot that Craven came up with did not represent the guess of any individual expert. Instead, it was the collective result of the independent analyses of a number of experts in different fields. Five months after the disappearance of the *Scorpio*, a navy ship found it 220 yards from the spot where the group collectively had said that it would be.[1]

The pattern formed by a large number of individual examples strongly suggest that this methodology should be investigated further for future problem solving, because it takes into account the experiences and views of a number of people. There are methods, which are based on the principle of collective problem solving, which have been tried and tested in controlled experiments and a number of practical analyses and forecasting situations. We will look into two of these methods here.

12.1   COLLABORATION CAFÉ

The Collaboration Café method represents a way of rapidly extracting individual knowledge from individuals, merging this with the knowledge and ideas of other people and making it into concrete ideas and structural capital. Based on research on effective group dialogue and social process, the café methodology took shape starting in the 1980s. Key proponents of this method are Peter Senge, the previously mentioned author of *The Fifth Discipline*, and his student Juanita Brown, an experienced conflict negotiator and group facilitator. Juanita Brown has developed a methodology she terms "The World Café." In 1999, Stuart Pledger, a friend of mine and a contributor of ideas during the writing of this book, created his own variation of the World Café, called Collaboration Café. Collaborative workshop methods have been widely used in business, dealing with complex issues.[2] The Collaboration Café format incorporates strategic questions, brainstorming for innovative solutions and a system for evaluating ideas produced by participants. The final phase of the Collaboration Café moves participants to commitment and activity planning for implementation.

The Collaboration Café method is built upon the idea of a "dialogue," which is held by a diverse group of participants. It has been used in order to develop creative solutions to challenges in groups of between 12 and 1000 people at the same time.

The idea behind the Collaboration Café methodology is to carry on a coordinated exploratory dialogue focused on questions of common interest. The dialogue is held in small groups of six people who, around café tables, share their findings and intermingle with other groups, also conversing on the same subject or on different aspects of the issue at the same time. The small groups may be located at the same venue or may participate via a direct Web broadcast.

The collaborative process is facilitated by a clear meeting structure and leadership complete with clear goals for the issues that will be explored, how the conversations will be held and finally in which form the output will be documented and communicated. These components greatly increase the efficiency and the quality of the conversations in the café sessions.

In the course of the Collaboration Café activities, which could be held during the energy transition program, a number of different groups could develop and refine scenarios for energy transformation of particular sectors, from now and into the future, and through "back-casting," identify the steps that are probable or necessary in order to arrive at the scenarios. Each Collaboration Café workshop, in a sequence, could focus on different themes, allowing for the gradual refinement of issues and scenarios.

12.1.1   Example of a Series of Collaboration Café Workshops

In this example we may assume that we are going to arrange a series of three workshops with people who will collaborate on the development of a strategy for the transformation of energy use in the transportation sector. The participants of these workshops will need to have some experience from different aspects of transportation.

Workshop 1 will focus on outlining transformation visions in transportation for a particular country.

Workshop 2 will work on a quantitative assessment of the potential of the visions for the same country and assessment of possible growth/decline curve for various energy solutions that are currently in existence and that are expected to be introduced in the future.

Workshop 3 would develop steps and measures (policy measures, innovation activities, etc.) that need to be taken in the country or by the companies in a particular industry in order to achieve the desired level of energy systems transformation.

There may be an interval of anything between a number of weeks and several months between each of the workshops. During the intervals, the results of the previous workshops have to be analyzed, complementary data could be collected through interviews and secondary sources, and, based on these results, the next workshop would be prepared. After the last workshop the results of the whole sequence could be analyzed and documented.

12.2   PREDICTION MARKETS

In the case of prediction markets, this tool is a more formalized way of applying the same type of thinking to complex forecasting situations. While the Collaboration Café method is ideal in aggregating the knowledge of a large number of workshop participants to an approximative result that is normally backed by all, or most, of the participants, the prediction market method can deliver very precise quantitative forecasts in situations with high uncertainty. This type of tool has been applied to predictions of the outcome of US presidential elections, where the method has consistently proven to be more accurate than traditional Gallup polls.[3] The results of the prediction market efforts consistently, as opposed to Gallup polls, correctly predicted the winner in each election. The prediction market methodology has also been used in predicting the risk of outbreak of war.[4]

In the case of predicting the outcome of new product launches, the prediction market approach has been used by the high-tech company HP in predicting the sales figures of new computer printers. The method was applied in order to predict the outcome of eight product launches and the result, when compared to the real outcome of the launches, turned out to be substantially more reliable than the result of traditional market studies and forecasts that were conducted for the same launches. The prediction market method showed substantially better results in six cases out of eight. The researchers also concluded that the results of these sessions are encouraging, because predictions consistently beat the official HP forecasts. One of the reasons why the method is of high interest, the researchers conclude, is that it is flexible and it makes it possible to aggregate any type of information possessed by different people: "It provides a natural methodology for quantifying subjective, qualitative, information giving weights to the opinion of different people for the purpose of information aggregation. The task is performed giving not only a point of forecast but also a complete probability over the range for which the value of some unknown variable is to be predicted."[5]

In the energy transition with a number of different opportunities that exist for each transformation area, prediction markets could be used in order to make detailed strategies and plans for the volumes of each particular type of fuel or technology that could be supplied or demanded over different time frames. I am indebted to Mikael Edholm, founder of Predicom, for the material in this section.

The prediction market method is based on the use of an electronic marketplace where participants trade event derivatives, that is, holdings in alternative outcomes of future events. Participants trade anonymously, using a fictitious currency, thus allowing the market mechanism to efficiently aggregate information and produce highly accurate forecasts. A prediction market will operate in a manner similar to a traditional "Delphi poll," but with many more iterations and the added benefit of producing real-time, continuous forecasts.

This method also allows the use of consecutive markets, making continuous refinement and increased validity of the resulting forecasts possible.

Box 12.1 How a Prediction Market Works

In a prediction market, the participants "invest" fictitious currency in different quantitative future outcomes. An exemplified prediction market for the transportation sector could consist of four alternative issues that participants could evaluate and "invest" money in. The issues may be

- the growth of alternative fuels;
- the growth in market share for hybrid vehicles;
- the increase in electronic commuting, reducing the need for personal transportation; and
- the increase in the number of car-free city centers.

Each of these issues may be set up as a set of binary investment alternatives. The growth of alternative fuels could be set up as two alternative investment alternatives in the following way:

1. By 2020 renewable fuels will command <30 percent of the fuel market in my country.
2. By 2020 renewable fuels will command >30 percent of the fuel market in my country.

The objective of the prediction market participants is to make a profit from their trading. This is achieved by buying low and selling high, pocketing the difference. Each participant in a prediction market is given an account with a set amount of a fictitious currency, say 1000 "prediction dollars" or P$. This amount may be freely invested in either outcome by placing orders on the market. Each order specifies the number of securities (called contracts) that the participant is willing to buy or sell and the amount he is willing to pay for each. The market mechanism conducts transactions once matching orders to buy are found. The guiding principle is first-come, first-served. Orders will be completely fulfilled whenever possible, but may be partially fulfilled when buy and sell orders specify differing amounts. Prediction markets limit orders to a price interval between 0 and 100, making the price of a transaction equal to the probability of the outcome becoming true. Thus P$0 equals 0 percent probability of an outcome becoming true, P$100 equals 100 percent probability of an outcome becoming true.

Let's say an individual believes the market for alternative fuels in his country will command 40 percent of the overall fuel market by 2020. Should he decide to invest in Outcome 2, "By 2020 renewable fuels will command >30 percent of the fuel market in my country," his investment would be made by placing an order to buy. Let's say he chooses to place an order to buy 10 contracts at P$2 each. As soon as another individual places an order to sell at least one contract at P$2 or less we would have a match and a transaction would occur. The transaction price equals the probability of the outcome becoming true.

Thus, a transaction price of P$2 would represent a 2 percent probability of this particular outcome becoming true. Since the prediction market was designed to be binary – having two opposing outcomes – this 2 percent probability of Outcome 2 becoming true also equals a 98 percent probability of outcome 1 becoming true. In the same way, an order to buy contracts in an outcome at any given price always equals an order to sell contracts in the opposing outcome at 100 minus this price.

Each participating individual may trade either outcome as often or as seldom as he pleases, basing his choice upon whatever information he finds relevant. This information could be external events, such as oil price changes, supply restrictions or regulations. Or, it could be information internal to the market – what might other participants do if the price changes? Will they buy, will they sell? Thus, each participant will take into account not only his own opinion of the validity of an outcome but also what he believes other participants will do, before placing his order. This is done over and over again, as participants and information find their way to the market. In this way the market will aggregate all available information, prices will rise and fall and the market produces a dynamic, continuous prediction of probability.

In the energy transition program, prediction markets could be used in order to make quantitative predictions of future outcomes, of the type indicated in the example above. Prediction market activities allow for the establishment of prediction markets for different energy technologies, and the potential and growth of each of them could be analyzed by any number of participating experts, who only need to have access to the Internet, in order to participate. As mentioned above, a number of prediction markets could be run in sequence, in order to gradually refine predictions and introduce new topics for analysis, based on the outcome of previous markets.

Financial Tools

It is probable that the steering committee, at the highest level of the program, will want to use different financial tools for different purposes within the program. Below, there is a list of a number of financial tools, which have been used in previous projects of different scope and scale, sometimes mentioning case examples of how they have been used and the results.

13.1 DIRECT INVESTMENT

In the Apollo program, the US government, through NASA, made a number of direct investments in facilities that were necessary for the success of the program. NASA invested in the facilities at Cape Canaveral and Kennedy Space Center. The construction of these facilities was necessary for the training of astronauts, launch of the rockets and the control of each space flight. NASA also invested in the development of all the technologies that were included in the rockets and other equipment that were used during the program.

The direct investments were made in the form of large-scale construction projects and in research projects of different kinds. Substantial investments in research projects have also been made in the general purpose technologies that have been studied by Professor Ruttan, described previously in this book.

13.2 TAXES

In order to support the growth of demand for renewable energy sources, many countries levy taxes on fossil fuels. A large share of renewable fuels in the mix for electricity production, 25 percent, has been achieved in the Netherlands. This unusually high share has been achieved because of high taxes on fossil fuels, which make green electricity comparable in terms of price with brown electricity. In the Netherlands, the market for green electricity opened before the market for brown electricity, which has also contributed to the high demand for green electricity.[1]

In an energy transition program, taxes, subsidies and price controls can be used in order to promote investments in selected renewable energy technologies, and to discourage companies to invest in nonrenewable technologies. Within a program, where the goal is to replace a certain amount of nonrenewable energy technologies with sustainable ones, these tools could be used in a systematic fashion in order to support both the investments in new technologies and the migration by consumers from one technology base to another.

13.3   SUBSIDIES

Many countries use subsidies in order to promote investments in renewable energy technologies. This may come in the form of subsidies for producers of energy, so that they invest in new technologies, or to subsidize the adoption of green energy by consumers. In Sweden, authorities have for a long time subsidized the implementation of heat pumps, wood pellet boilers, and improved insulation, in order to reduce energy consumption in the cold northern climate.

13.4   PRICE CONTROLS

A measure that is similar to subsidies is price controls, through which the government guarantees that utilities get a minimum price for its green energy. This reduces the risk for investors and makes larger amounts of capital available, at a lower cost, for the investments in green energy technologies.

13.5   MARKETING OF AND INFORMATION ABOUT RENEWABLE AND GREEN ENERGY

In a market economy, free markets and pricing do not solve every problem, and the lack of demand for energy produced from renewable sources may in some cases be accounted for by insufficient marketing and information efforts, rather than a lack of interest among customers. In the case of the Netherlands, where 25 percent of all electricity is bought in the form of green electricity, the rollout of the green electricity market, the pricing and the marketing efforts have supported the large-scale adoption of green alternatives. Contrary to this, the rollout of the green electricity market in the UK has largely failed, despite seemingly favorable basic conditions for the large-scale production of competitively priced green energy. In his book *Renewable Energy Policy*, Paul Komor identifies three main reasons for the failure of green electricity in the UK:

1. Consumers focused on cost as the main driver for switching. The cost focus is largely due to the way that the electricity market was presented in the UK – as a way to reduce the cost of energy.
2. "Lousy marketing by incumbents," which means that utilities did very little in order to promote green energy alternatives to their

customers. One company only wrote about it briefly in its customer loyalty magazine.
3. Uncertainty about the way that the electricity market, and the market for green electricity in particular, was going to work made the companies cautious about investing in renewable energy sources and to promote green alternatives.[2]

This example from the UK indicates that markets work well when they are managed well. Markets do not organize themselves. Governments need to involve themselves in the planning and management of energy markets and markets for energy technologies and green products.

13.6 MAKING ACTUAL COST SEEN AND FELT

In many cases, competition between different alternatives is biased because some of the costs of one alternative may be paid through taxes, or paid for in other ways that are not seen or felt by the consumer at the point of a decision about different alternatives. In the competition between cars and trains, the cost of building and maintaining roads is, in most countries, paid for by the public sector, via taxes. This means that the cost of building and maintaining the road system will seldom be weighed into an equation when drivers decide whether to drive, or to take the train. In the same way bus systems, and truck transportation, are subsidized as well, and operate at a cost advantage, compared to trains and airplanes, as long as buses and trucks do not have to pay extra for the mileage that they do on the roads.

In some countries in Europe the motorway systems are toll-based, and in Germany foreign trucks have to pay a fee per mile for using the road network. This system of payments is administrated through a network of fixed high-tech registration units that read identification devices in trucks, as they pass certain locations along the road network. In some cities, Oslo and Stockholm for example, fees have to be paid by all cars that enter the city centers.

Increasingly, high-tech solutions become available, which can register the vehicles that pass through a certain registration unit along a road. In this way, it becomes not only possible, but also practical, to let drivers, or transportation companies, pay for the actual mileage that they do on the road network, or do the same for certain parts of the network such as large cities or motorways, or for categories of vehicles.

13.7 COMPLEMENTARY CURRENCIES

The least well known of the financial tools in this listing is complementary currencies. This is a tool that could take a number of forms. The purpose is to make resources, usually in the form of time, available for activities, which could not easily be financed within the national currency systems. Complementary currencies could be used in the form of complete currency systems, which exist in parallel with national currencies. Complementary currency systems could also be equipped with mechanisms that are different from national currencies. One of the advocates of complementary currency systems, Bernard Lietaer, formerly a senior manager at the central bank of Belgium, has written extensively about complementary currencies in his books *The Future of Money* and *Of Human Wealth – Beyond Greed and Scarcity*. Another author in this field is Thomas Greco, author of the book *Money*.

The primary mechanisms of national currencies are interest rates and mechanisms to establish exchange rates, which are primarily market-based. Earlier currency systems involved the gold standard, which has gradually been abandoned in favour of free floating rates of exchange. The mechanisms create a scarcity of currency. Different mechanisms contained in complementary currency systems could contribute to eliminate scarcity and make room for economic transactions and growth that would not be possible within the realms of our existing national currencies. The mentioned authors describe in detail how these types of systems work.

It is conceivable that an energy transformation program could require a number of different activities, which could not easily be financed through national currencies. If it is decided that government should finance the energy transformation program, the financial burden of the transition may be heavy. It could, in this case, be an advantage to use nationally sanctioned systems of complementary currencies in order to relieve the national currency systems of some of the financial burden of the change. The organization of these systems could take many different forms. One thing that we can be certain of, however, is that they need to be well planned and managed in order to work well.

Box 13.1 Taking Care of the Elderly in Japan

The past decades have been difficult in Japan. During more than a decade, economic growth has been low or negative, and many people have been unemployed. These economic difficulties have given rise to a large number of complementary currency systems, which have helped in the creation of new markets for goods and services.[3] Complementary currency systems can be equipped with mechanisms, which make it possible for people who want to do business with each other, to exchange goods and services for complementary currencies, and "create" the money through the transaction. (In a national currency system, new money is created by banks when they lend money to customers.)

In Japan a complementary currency system has been in place on a national basis for a number of years, which serves to create services for the elderly. Many people live a long distance from their elderly parents, and find it difficult to visit and take care of them in the way that they would have liked to. In order to be able to buy time from other people, who could take care of their own parents, it is possible for people to earn money in a complementary currency system, by taking care of someone else's parents. Each hour spent by a person, taking care of elderly people, will earn the same person a 1-hour equivalent in the complementary currency. This could be used in order to pay someone else for taking care of one's own parents where they live.

Every participant in the system is free to decide how much time he or she is willing to spend, earning money in the system. A person, whose parents are ill, and need a lot of care, may spend a lot of time taking care of elderly people, in order to make many hours of care available to his or her parents, in turn. A person, who has parents in less demanding circumstances, could spend less time and earn less money.

The advantage of this type of currency is that a person, who is unemployed, or poor, will not have to spend precious money, in yen, in order to pay someone to take care of their parents. There is also no scarcity of currency in this system. Every time someone decides to spend an hour taking care of an elderly person, she can earn the time equivalent in the complementary currency. In this way, valuable time is made available for services to the elderly, which would otherwise have been spent on less productive activities.[4]

In the energy transformation program, there may be a multitude of tasks that would need to be performed, which may be difficult to pay for through national currencies. We have mentioned earlier the need for widespread information and training activities. Some such activities could be organized as neighborhood projects, or work groups, through which families could get information on how to reduce energy consumption. It may be difficult to get people to attend such meetings, and to motivate them to change parts of their lifestyles or consumption habits. Complementary currencies could be used in order to reward people for attending meetings, and they could also be used in order to reward lifestyle changes, investments in energy-efficient equipment or people who take community initiatives to conserve energy.

In such a system, we could think of a number of situations, in which people may be allowed to spend the money that they have earned. Some complementary currency systems allow members to use the currencies to, partly or wholly, pay for goods and services in regular shops. In such a case it is important to devise a way for shop owners to redeem the value of the currencies through exchange in their turn. It is also conceivable that complementary currencies could be used in order to pay for tickets, food or services at events that are created for the benefit of participants. This could be festivals, entertainment or other types of events.

Complementary currencies, used for different purposes, could be used as potent tools to drive change.

Box 13.2 Environmental Work in Curitiba

The mayor of the Brazilian city of Curitiba wanted to do something about the garbage that was polluting the streets of the city. In order to achieve this, a complementary currency was implemented, which people earn by gathering waste and depositing it at waste stations around the city. The money earned through these activities can be used as payment on the local transportation system of Curitiba. This has the added benefit that many poor and unemployed inhabitants of the city earn money, so that they are able to go to town and apply for work.[5]

CHAPTER 14

Change Happens in Steps

Contrary to what many people intuitively tend to believe, change usually happens in steps, which may be more or less distinct. Yet, we sometimes perceive change to be radical and sometimes surprising. It may seem as if a new technology, or product, has been developed, launched and spread in almost no time at all. This, however, is never the case. Technologies always develop in steps, and to experts in an area, a particular development is seldom surprising. A new technology is generally developed and spread at a low level before it becomes widely available to everybody in the open market.

Box 14.1 The Rapid and "Unexpected" Growth of the Internet

To many people it may have seemed as if the Internet and its related technologies did arise out of nowhere in almost no time at all. This is not the case. The electronic transmission of messages between computers was foreseen by many experts for a long time and the technologies that make this possible were developed in a number of steps.

Ever since the invention of the computer, the advantages of direct communication between computers have been obvious. This is because cooperation between partners in different companies that use different computer systems (this could be business systems, or systems for document archiving or retrieval, calculations or other computerized applications) often involves the exchange of data. When these data need to be exchanged on paper, or sent by mail between partners, the cost and time spent on the exchange of data are often substantial, because there will be duplication of work in entering the data two times into different computer systems. When computers can exchange data automatically, without the need of people as interfaces to transfer data between the systems, the cost of exchange of data is dramatically reduced.

A number of different developments form the background of the Internet and the technologies that are necessary in order to facilitate

---

electronic communication. The first electronic network was the ARPA-NET, which was started in the late 1960s to connect computers at a number of research facilities across the American continent. The idea of the ARPANET was developed by Joseph Licklider, Director of the ARPA Information Processing Techniques Office, and the first network connected 15 major research centres from the East to the West Coast.

In order for the network to grow and allow increasing numbers of computers to be individually identified, new transmission protocols had to be developed. In 1973 Robert Kahn together with Vincent Cerf developed the transmission control protocol (TCP), on which Cerf and Robert Metcalf had earlier collaborated. This is still the type of protocol that is used for Internet communication. After this a number of further innovations have made the present Internet possible. However, the emergence of the public Internet in the 1990s was not a surprise to insiders in the computer industry. The vast consequences of this development, however, were still in 1998 foreseen by few.

Gradually, the idea of sending electronic messages and other data across networks that would be open to the public spread and the opportunities were discussed and analyzed for many years. The development of electronic data interchange (EDI), which was a precursor to Internet-based information exchange, happened in parallel and indicated the substantial opportunities for cost reduction that were embodied in electronic and automatic exchange of information between computers. With the Internet a set of technologies were developed and integrated that facilitated the cost-efficient implementation and use of these opportunities. The ideas that underpin the Internet, however, had been around for a long time.

Despite the many examples of new technologies, and the social change that results from this, many people believe that the future will primarily provide us with more of the same, rather than something entirely different. As people now discuss the need to reduce energy consumption, many accept that we will need to cut down on our use of energy. Yet, most of us still find it difficult to envision the changes that this may require, and find it especially difficult to grasp that we may need to change our lifestyles in order to save energy. Since companies have never cared about energy in the past, it is also difficult to accept that companies will need to care about energy in the future. These are only two examples, among many, of the difficulties that are involved in understanding change.

Box 14.2 The Combination of Large-Scale Investments and Lifestyle Change

One of the reasons why it is difficult to grasp the nature of future change may be that the means of change are not always at our disposal when people start to discuss change. For example, the reduction in energy consumption may seem impossible to accomplish for individuals, who need to use their cars to travel to and from work, and who live in houses that require certain amounts of heating or cooling. How are we supposed to dramatically reduce our energy consumption to the sustainable levels that environmental experts discuss?

In order to achieve a dramatic reduction in energy consumption and carbon dioxide emissions, we will need large-scale investments in new energy systems, and transportation solutions as well as increased use of new energy-saving technologies by individual households and companies. If we want many more people to use energy-efficient transportation systems, or install heating based on renewable energy sources in their homes, we need to first make the new systems available at an affordable price and then market them to users. Some of the alternatives for investments are described in Part III.

As a comparison of another recent change that has been based on a combination of large-scale investments and changes in consumption patterns and work life change, we could take the increased use of electronic communication. The rapid growth of the Internet, and the computerization of homes and offices, has been achieved through a combination of an increasing supply of inexpensive bandwidth, PCs that are available at shrinking prices and a large number of electronic services that have become increasingly user-friendly. This has caused an increased demand for PCs and new services, as they have become available, which has, in its turn, made electronic services available to increasing numbers of people.

The difficulty in forseeing change may also exist because we understand the logic of existing trends and we find it difficult to understand an emerging logic that may radically change society or businesses in the future. A very large number of people now understand how to achieve operational efficiency in companies, and many believe that the improvement of operations and the structuring of markets are trends that will continue forever into the future. As I have described earlier, the development of operational efficiency has taken a long time and it has seldom been obvious to people at the start of the project that the surprisingly large improvements that are later realized would even be possible. The idea that a new development may force us to reevaluate a number of our present business practices and aspects of our lifestyles, at this point, seems farfetched and abstract to many people.

Since this book challenges the belief, that existing trends will continue, we need to identify new trends that can replace the existing patterns and we need to identify and understand the reasons why these trends could become predominant in forming the future. As we have argued above, we can identify the new force of sustainability as perhaps the predominant business trend of the future. The reasons for this are the two facts that oil may become scarce and increasingly expensive in the years to come and that pollution and climate change are beginning to cause large-scale problems, which may affect the conditions for life for humanity and other species on this planet in the near future.

Another very telling example of how we fail to foresee imminent changes in society, the author of this book was among the early analysts of the emerging e-business boom. Together with colleague David Lundberg, we published an analysis of the strategic consequences of electronic communication and business, in the book *The Transparent Market*, which was published in 1998. At this point, only 2 years before the "Internet boom," very few people, including managers and management consultants, could understand the impact that the Internet would have on the ways that companies communicated and did business. We argued that not only would companies in the future send orders over the Internet. All kinds of business information would be exchanged between business partners and companies that searched for suppliers and those that used the Internet as a marketing channel to new customers.

Here some of the trends toward sustainability will be identified, and the steps that society will have to pass through from the initial stages of awakening some interest among a few, to a full-blown focus and paradigm shift in terms of our use of energy, will be identified. In this book I argue that we will experience change in the near future, that the nature of this change can be analyzed and understood in advance and that measures can be taken to help us enjoy the advantages and avoid some of the disadvantages of change.

In case governments decide for planned change, the different steps could be made into parts of the plan, and the steps could be milestones in the development toward a sustainable society.

14.1 THE E-STEP CHANGE MODEL

Based on the insight that change happens in steps, David Lundberg and I in 1999 developed a model through which business transformation in various industries and sectors of the economy could be analyzed. When the step change model was developed, the Internet in general and electronic business in particular were in their infancy. Most large companies already had Internet sites, but very little change had occurred in their business models. At the same time there were entirely new businesses that had been developed on the Internet, but still only few consumers used electronic services. Amazon had been around for 3 years, but few people outside of the United States had used the service. Still Amazon had already served 1 million individual customers, but the company was operating at a huge loss. Internet banks had started to grow, and the Swedish bank SEB counted as the largest one globally in terms of the number of users, even though this bank is based in a country of only 9 million people and the bank at this time did little business outside of its core market. Many business people, at this time, believed that electronic business would always remain a marginal phenomenon, and that companies that do electronic business, similar to Amazon, would always have to struggle to turn a profit.

In order to explain why there was so much hype around electronic business and the changes projected by the proponents, despite the fact that so little had changed even though virtually all companies were already online in some form, we developed the e-step model. This model is based on the ideas that change happens in steps and that it is possible to foresee which steps a particular development is going to pass through. The prerequisite for the ability to forecast the steps in the change is that a sound analysis needs to be made, through which the analyst can identify the drivers behind the change and a number of other variables such as opportunities to reduce cost and the opportunities to create new or increased value offered by technology. We did the analysis ourselves with very little input from other sources. It is reasonable to assume that an even better and more detailed analysis could have been made using the Collaboration Café methodology or prediction markets, had they been available to us at the time or that we could have arrived at more detailed predictions if we had the resources to do more data gathering, interviews and analysis. Different methods could be used on their own or in combination, depending on the situation and the issues that need to be analyzed.

We later developed a general model for the development toward electronic business, which was first published in a book in Swedish[1] in 2000. Later still, we analyzed and forecast the stepwise development of e-business in a number of business sectors and industries and found out that each particular situation was unique and that e-business would develop in different ways in each sector that was studied. Regardless of this, we still find that the e-step model describes the general development steps of e-business in such a way that it enhances the understanding of business change. This is because it emphasizes that business change is a cumulative process, in the way depicted in the model of organizational learning, which was presented in an earlier chapter. The cycle of learning that is described in the model for organizational learning is reiterated several times through the e-step change model, so that at each step and perhaps more than once for each step, the cycle is gone through, individual learning is formalized and made into structural capital and turned by companies into processes and routines that are really being used.

I will now briefly go through the development steps of e-business as they were described in the original model. This will be done in order to familiarize the reader with the step change model and first use it in order to describe the steps of e-business change that are familiar to most of us. After this, the energy step change model will be described below in its general form. In each particular business sector or industry, which will need to be changed during the transition, the steps will be different. This is to some extent indicated in Part III of this book, in which the present situations in different sectors are outlined from a change management perspective. Again, however, the general model will assist the reader in understanding the progress of change and the fact that not very much is changed during the first few steps. Despite this, at some point in the near future, the development will reach the point where the slowly growing interest in energy issues will become replaced by rapidly growing interest and attention, both among companies and households.

First, we take the step-by-step development of e-business, as it was predicted in our original model, published in 2000.

14.1.1   E-Step 1: E-Windows

The first step in the development of electronic business was the creation of noninteractive company sites on the Internet. These sites contained some information about the company's products and services and some contact information. In 1999 they rarely contained prices and it was usually not possible to place orders or interact with the company in any other way. Yet, people that we met kept asking us how we could argue that the Internet would be able to change the way that companies did business. Most companies already had sites and this had not changed anything. The answer was that e-windows only constitute the first step, and an electronic shop window or an Internet-based brochure changes very little.

14.1.2 E-Step 2: E-Channels

In 1999 some companies already had sites on which it was possible to order goods and services. At this time, however, these sites tended to be mere electronic channels for transmitting order information to a traditional store, or an Internet-based store, that was operated in the same way that all other physical stores were. The main difference that was created was that orders could be transmitted electronically and that products were picked, packed and sent by employees of the store, instead of being picked up from the shelf by the customer. Many times this manually delivered service only added cost to the price ticket and it was hard to drive the argument home that over time and in many cases, e-business would save money for customers, through reduced prices. Examples of this type of e-business solution was (and still is) when a grocery store collects orders electronically and when employees pick, pack and send groceries to customers from the store. Almost nothing has changed in the business model and cost is often added rather than reduced.

14.1.3 E-Step 3: E-Breakouts

Early in the development of the Internet, some companies created business models that differed so much from traditional business models in their industry that they represented a new way of doing business. This is because the electronic business alters the cost structure and allows the company to serve its customers in a new way, adding new value and decreasing price. An e-breakout could be a breakout from the existing structure of the company that started an electronic business or it could be a start-up company that creates a breakout from the traditional structure of the industry.

An example of this would be Amazon.com. Traditional bookstores need to have one store in every city or one in every part of the city to serve its customers. Amazon runs a few bricks-and-mortar distribution centers, but the company greatly reduces the cost of storing books, records and other products in hundreds of locations across the country, which is the case for ordinary bookshops. Amazon is also able to reduce the number of employees, compared to traditional bookstores, since it is operating out of only a few locations worldwide. In terms of customer service, the electronic services of Amazon can add new value to customers by recommending books for them online and they can also offer millions of books, in fact virtually all books that are available in the English-speaking book market, and present updated information about the contents and the availability of each and every one of them.

Another example, which reduces the cost even more radically, is the Internet bank. According to a widely cited example, a transaction performed in a traditional bank costs the bank 2 dollars, while a transaction that is made through an Internet bank costs approximately 2 cents. This is because the Internet bank does not need to keep offices and there are no employees involved in the transactions. At a traditional bank office a transaction takes a few minutes, and this time, on average, costs 2 dollars, everything included. Even though the services offered are the same as before, electronic business has changed the way that the services are delivered and reduced the cost of delivering standardized services.

At this point it becomes clear that, if many companies apply advanced electronic business models, it would dramatically change the business landscape. Conversely, if many customers use the electronic business services that are available, traditional stores and other services will soon be forced out of business. In the book trade, Amazon, reportedly, now has 10 percent of the US market, but the total market has grown in the past decade since Amazon was started, so there is still substantial business left for bookstores. In the case of music stores, the electronic downloading of music has hit the stores and the music companies hard and in many cities and countries specialized music stores are closing down or being transformed. Since electronic business is still growing year by year, we have yet to wait for assessing how much change the Internet will bring to the retailing landscape.

In business-to-business situations electronic business models have become much more commonplace, and EDI in very advanced forms has replaced transmission of orders via phone and fax. Instead, computers at purchasing companies now often communicate directly with computers at the selling company and orders are placed electronically without a manual interface. In this case, the change is less visible, because these companies did not have stores in the first place. They usually deliver goods from their storage facilities or sometimes produce to order and ship. In this case electronic interaction has dramatically reduced cost and it has often also reduced delivery times and other lead times in the process. In addition to this, e-business often makes it possible to dramatically reduce stock levels.

14.1.4   E-Step 4: E-Platforms

In the case of e-platforms not only the way of delivering the service has changed, but the whole product or service has been changed as well. In the case of music, this is no longer delivered on a physical CD. It is many times downloaded directly over the Internet. Many people also legally purchase music in this way via iTunes and a host of other services. The same is true for films that are distributed directly over a cable or broadband network. If books are sold as electronic documents and read on electronic devices, which have been available for some time, this would represent another example of e-platforms. E-platforms are easier to create for information products and services that can be delivered electronically over the Internet. Most physical products are difficult to change dramatically through electronic business and thus e-platforms may not be as relevant a step for those. Needless to say, in 1999 there were not many, if any, examples of e-platforms, and it was difficult to explain to people what these would be like and how they would change the behavior of future Internet users.

14.1.5   E-Step 5: E-Values

Also, in 1999, there were no examples of e-values and there are still not very many. This is the ultimate example of electronic business and it means that customers can use the Internet to configure products and services that entirely suit their preferences. We could think of a customer in the future who wants to purchase a book. The book is available in electronic form. The customer can decide if she wants to download the book in electronic form or if she wants to have it printed and bound on demand, using the paper and the binding that she chooses. She could then also have it delivered by the transportation company of her choice or pick it up herself at a delivery station en route from work, where she can also pick up groceries and other goods from different suppliers at the same time.

The customer who wants to do grocery shopping over the Internet could enter the shopping list online and ask the electronic service to determine where to buy different goods on the list and select the criterion to use. The customer may prefer ecological groceries at the best possible price and may ask the service to divide the order between the suppliers that together best meet the criteria. This may mean that bread is delivered by an ecological bakery and that the other goods except meat are delivered by an ecological food store, while the meat is delivered directly from an ecological farmer.

This brief description has hopefully helped the reader understand the purpose of the step change model, so that when we now apply it to energy transition steps, these will become comprehensible in the light of the e-business change that is already underway. The e-step model was developed for electronic business almost 10 years ago and this development was then perhaps at the level at which the energy transformation one is today.

CHAPTER 15

A Step Change Model for Energy

In the case of energy transition, we believe that the steps could be described as the following types of changes:

1. Low-level experiments
2. Change in values and mental models
3. Technology change
4. Incentive change and change in consumer behavior
5. Systemic change
6. Ecology change

15.1 LOW-LEVEL EXPERIMENTS

There are already people and companies that are interested in changing their habits with regard to energy consumption. Some of them find ways to reduce their energy consumption, through their own interest in the matter. Companies may have substantial resources to invest and pioneers may gain market share or improve profitability by reducing their energy consumption. Individuals could similarly make efforts to reduce energy consumption or switch to renewable energy sources. Richard Heinberg, author of four books on the peak oil issue, describes how he collects vegetable frying oil from restaurants, and how he uses this to fuel his car. The availability of this type of oil, however, is limited, and this model cannot be used for large-scale energy transformation.

In the same way, some people install new heating systems, or photovoltaic panels, at home, in order to reduce their energy consumption. These types of efforts are, naturally, important in order to test new technologies and to signal that change is possible, but they will not, in their present small-scale form and high cost in terms of money and time, solve the problem of energy transformation. In order to do this, we need to create large-scale cost-effective solutions and initiate change broadly in society. In many cases this requires large-scale investments in new infrastructure or large information campaigns to inform many people about opportunities to change behavior or make investments in energy-efficient products and technology. Before demand is created, however, a sufficient supply needs to be secured and the supply of existing alternatives needs to be reduced. In order to transfer 15 percent of European vehicles to biofuels we would, for example, need some 150 large plants that produce these biofuels and the large-scale logistics systems that supply them with raw materials and distribute the fuels to the market. In the absence of any large-scale production and distribution facilities for alternatives we need to settle for low-level transformation activities.

15.2 CHANGE IN VALUES AND MENTAL MODELS

We behave in certain ways because of our values. If we want to change behavior, we need first to change the values and mental models of people. In order to illustrate this, we may think of how slavery was once abolished in America and other countries. According to John Steele Gordon and other historians, it was not seen as a moral issue in seventeenth-century America to keep Blacks as slaves. People of African origin were seen as inferior to the Whites and this view was seldom questioned. In the eighteenth century, people were influenced by French philosophers, who argued that all human beings had the same value and the same rights. Values started to change and even many rich people, such as Benjamin Franklin, who held slaves started to question their own right to do so. Initially, they did not know how to go about abolishing slavery, because it was a very important factor in the economy of the American South and the economy was dependent on slave labor. Over time, however, people found ways to overcome the obstacles, and they even found that freeing the slaves created economic advantages from a number of perspectives.

In the same way, a large part of the population of the Western world has seen it as their right to exploit the environment. Now, with climate change and other environmental problems, we find that we may not be able to do this for very much longer. As oil is becoming scarce, we will find that there may not be enough of it in the future, for us to consume it in the way that we have done in the past.

Many of us now realize that we need to change our habits. As we do this, our general values toward energy consumption are changing in favor of energy conservation, but we do not know what exactly to change. Most of us would probably like to change as little as possible in our lifestyles and then choose to change those things that have the largest impact on energy consumption and on the environment. If companies, or the government, could act to solve the problem for us, many people would probably embrace this as a perfect solution.

The ambition to focus on the changes that are least inconvenient or costly, and deliver the largest result, is also the soundest strategy from the perspective of society. Seen from this perspective, we also need to take into account the effects of energy transformation on business and the economy. In principle, even if some radical environmentalists would disagree with this, we would not want extremely dramatic change in the whole global population in one go, even if this was possible. The economy and the different players in the form of companies, workers, financial institutions and a number of other aspects of this system need to adjust gradually and in a planned and structured way. We need to develop a strategy for society, by which business and economic systems are able to gradually adjust to change.

Therefore, it is not crucial that all individuals completely stop doing certain things. It is good enough for the short term if we reduce the total amount of certain activities, which consume much energy and that are harmful to the environment. We do not know at present exactly which activities we need to change, how much we should change each activity and which pace of change we ought to go for. As consumers, we could use some advice on this in order to manage our own change processes and the same is true for companies that are also faced with a number of different alternative changes that could be made. In the meantime, those of us who are aware that we need to reduce energy consumption, could use available advice and do the best that we can. Among activities that we, as consumers, could reduce in order to make a rapid impact on energy consumption and CO_2 emissions, we find

- red meat consumption;
- purchasing of imported goods that have been transported by air or long distance by truck;
- driving, using cars with high fuel consumption;
- leisure and business travel by car or airplane;
- heating and cooling of houses and apartments more than necessary; and
- building new fuel-inefficient buildings and keeping old ones in a fuel-inefficient state.

We may want to change our values in a direction that helps people to reduce energy consumption in the following ways:

- Instead of wanting to eat a lot of red meat and large steaks, we can adopt the value that red meat is good on certain occasions, but that it is much better if we reduce red meat consumption by 50 percent and instead increase consumption of vegetables, fish and poultry, which require far less energy to produce. We could even reduce meat consumption altogether by 50 percent, choosing dishes with less meat per serving, such as pasta, Thai or other food with a lot of vegetables, instead of a steak, chicken breast or tuna steak.
- Instead of increasing our leisure travel by going on an increasing number of short trips, we may instead choose to go on fewer trips of longer duration, reducing the total air mileage for leisure travel.

- Instead of keeping an SUV in the driveway, we may choose a more fuel-efficient car. This could be a smaller gasoline or diesel car, an electric car or a car with a hybrid engine. For those who need to go on driving a big car, they could try to reduce the amount of driving, through better planning of shopping trips, or through carpooling.
- Instead of heating and cooling houses to keep the desired temperature in all rooms all the time, it is possible to adjust temperature so that the desired temperature is kept when people are at home and that heating or cooling is reduced in rooms where people spend less time and at hours when people are not at home at all. Initially, this may have to be done manually, but if people demand such facilities, automatic control systems can be installed in houses and apartments when they are built or refurbished.
- Instead of building houses using inexpensive building materials or keeping houses that are fuel-inefficient, we could build houses that conserve energy. This could mean that houses are built as "passive houses" so that they stay cool for longer, even when air conditioning is turned off. In colder climates it may mean that houses need less heating in winter, because of construction solutions that help walls and floors to accumulate heat during the day, use better insulation in walls and windows that are better insulated as well, using triple glazing or similar techniques.
- Instead of indulging in exotic food from other parts of the continent or the world, people could increase consumption of locally produced vegetables, beverages and other food and eat and drink less of the stuff that needs long-distance transportation.

The above are examples of possible changes in specific values and the changes in behavior that may follow from such a value change. In total we could save substantially on our energy consumption in the United States and Europe if everybody did the above. To achieve this, we would need a lot of information, advertising and, probably, also incentives. If we wait until energy prices increase to the level at which people can no longer afford their high levels of consumption, our economy may already be in dire trouble. I will treat behavior change as one of the possible areas of energy improvement in the last part of this book, which treats the program structure.

If everybody would change a little in the above ways, we would collectively save a lot of energy. This would buy us time to develop more and better technologies that are based on renewable fuels, energy-efficient systems of different kinds and also economic and financial solutions that could carry the investment needs and it would send clear signals to business that values are changing in favor of more energy-efficient lifestyles, products and industrial processes.

15.2.1 Individual and Corporate Values

As we have seen in the model of organizational learning above, corporate values and corporate behavior are influenced by individual knowledge and values and individual behavior. If people change their values and behavior at home, we must expect them to bring their new values and behavior with them to work as well. We would also expect the process to work in the reverse fashion, so that new values learned at work will to some extent translate into changed behavior at home.

Using the model for organizational learning, we may expect people to start to think through the way that the company that they work for operates and bring some of their tacit knowledge about company energy consumption out into the open and make it explicit. People who reduce their leisure travel or their driving through better planning or trips of longer duration may want to do the same when they travel for business. They may try to meet more people in one trip, instead of going several times, or they may use video conferences or phone calls in between personal visits to remote offices and customers, instead of traveling as often as today. As managers describe their new values regarding travel to their fellow employees, this would be an example of how individual tacit knowledge became explicit and it would then constitute a unit of organizational tacit knowledge, which rests with the individuals who have changed their habits, but not yet made it into corporate practice and corporate structural capital.

The individual who successfully tries to do her work using less energy may tell colleagues about this and suggest that they do the same. As more people take on the habits of planning their traveling and use IT for communication, the organizational tacit knowledge will become explicit and formulated into communicable principles for business travel. At an early point they will exist only in spoken form and they will be communicated by word of mouth.

As more people in the company become interested in energy-related issues, these principles will become formulated into written operative instructions, even into strategy documents and other explicit forms of organizational knowledge, which make up its structural capital. As they are communicated and used by the whole organization, they will become practices being used throughout the organization. These processes could be sped up by companies or individual employees who take initiatives to gather employees in meetings, perhaps run Collaboration Café sessions and identify energy savings opportunities collectively that could rapidly be made into corporate practice.

The examples from other companies that develop new technology, change work practices or implement new products or technologies in their businesses, presented in this book and other books, are examples of how learning can be spread between organizations. This book is read by individuals, who bring it as tacit organizational knowledge into their organizations, where it may gradually become explicit, written into documents and taken into the structural capital of the company and taken into use by the organization.

In the same way that there are improvements for households and individuals that can be made immediately or after a relatively brief analysis, there are also improvements in companies that can be made relatively rapidly. The problem in a company will be that there are a large number of possibilities and that they need to be spaced out in time in a plan, because of the limited resources for investments, project management and for technical installations, testing and training on new installations. Each opportunity of the total number of opportunities in a company needs to be prioritized and scheduled for completion as part of the overall plan. Here are a few examples of improvements with a short lead time and low-investment needs:

– Reduce business travel through video conferences, phone conversations and better planning of trips.
– Improve energy efficiency of pumps and other installations with frequency control of equipment.
– Improve energy efficiency of pneumatic systems by installing the new product "X-Block" offered by the company X-Design (See Box 15.1).
– Install sound and movement sensors that turn on lights when a person is present in a room or in a staircase and turn them off again when they leave.
– Install automation or sensors to control lighting, heating and cooling of office space and other areas, so that energy is saved in the evening and night, when nobody is present.
– Replace old heat exchangers, cooling equipment and machinery by new and more energy-efficient models.

In the past, the above types of investment had not been prioritized. It has been taken for granted that successful companies can make enough money to make up for a certain amount of waste. Getting people to save energy has almost been at the same level of priority as withdrawing free coffee for employees. With a change in individual values among employees, the tacit knowledge in the organization regarding where there are energy savings opportunities in the business may be formalized and turned into structural capital. This structural capital could include values regarding energy savings and changes in processes and work practices so that energy savings are realized.

Box 15.1 Save 50–70 Percent on Energy for Pneumatics

Pneumatic systems in factories, which are used as power supply to a number of pneumatically powered tools and machines, are notoriously energy-inefficient. This inefficiency is caused by a number of factors. Among them, leakage in the system is an important source of inefficiency, but the way that pneumatics is run, making use of pressurized air only once, and then emitting it into the air of the surrounding plant, with the resulting need to constantly re-pressurize the system to the necessary working level, is another factor.

In order to solve this problem, an inventor and entrepreneur in northern Sweden, Per-Åke Wahlberg, has invented a pneumatic block, which keeps the pressurized air inside the pneumatic system, using it to power the return stroke as well as the forward stroke. This reduces the need to build new pressure, and it keeps the pressure of the system at a more constant level. This not only has the advantage of reducing energy consumption for pneumatics by 50–70 percent, but this also makes the whole plant run smoother, reduces vibrations dramatically and thus reduces the machine breakdowns and other maintenance and production costs.

The block is now successfully marketed by the company X-Design and this company has started to launch the product on a global basis. A number of large companies have signed up for pilot installations and full-scale runs. This is an example of how relatively low-tech innovations in systems that have been in use for a long time can contribute substantially to energy efficiency. The installation of the X-Block in a plant pays back in a few months, both through the reduced cost of energy, and through the reduced cost of maintenance and production breakdowns, that are achieved through reduced vibrations in the plant.

Solutions such as the X-Block are available in different areas. Many times improvements that are possible never get done, because people prioritize other things. This is often because there are incentives in place that make them prioritize in the way they do. When a change in values related to energy efficiency is on its way, some changes may be made voluntarily by some people, but even more changes will be made if incentives make people prioritize differently.

15.3 TECHNOLOGY CHANGE

Throughout the energy transition, we will need to develop and apply a large number of new technologies on a large scale.

We will need to replace many old and inefficient technologies by new and energy-efficient ones. This change may be the most difficult part of the transformation. It will be difficult because it requires huge investments in total, and investments require financial resources, that are, per definition, limited. Technology change is also problematic for another reason. It takes time. It takes time, not only to make the money available, it also takes time to make new technologies available and it takes time to make new products that are based on the new technologies available in the amounts that are needed for a rapid transformation. This requires the expansion of production capacity to the levels that will be necessary in order to ascertain rapid enough market penetration by energy-efficient technologies.

Box 15.2 The Timeline for Technology Development

Depending on the level of innovation that is needed, the development of new technologies, and the increase in production capacity of new technologies, will take different amounts of time. Professor Ruttan describes how large-scale government projects are necessary for the development of new general purpose technologies, because of the large investments and the long time that it takes from the start of this development to the commercialization of the resulting technologies. In the case of general purpose technologies, such as computers, nuclear energy, space technologies or the Internet, we often have to allow for several decades for this development.

In the case of the application of an existing general purpose technology in some new application area, we may expect the development to take a number of years. This would be the case for the adaptation of the internal combustion engine to new fuels, such as biogas, dimethyl ether (DME), or some other new fuel.

After a technology has been developed, it will take a number of years to retool plants in order to increase production capacity for the new products. Initially, ethanol-fueled cars, hybrids or electric cars will be available only in small volumes, while gasoline-fueled cars will still constitute the majority of the cars that are sold.

Another example of a development that has been discussed for some time is the development of advanced return logistics and "reverse factories." The car manufacturer Fiat has developed a concept car that can be automatically disassembled into pieces by robots and the parts can be reused in new cars or molten down and made into new parts. The large-scale application of this idea would require completely new design procedures, materials and methods and materials for assembling products. Seemingly, reverse factories and logistics systems add cost to products, but in large-scale and well-managed systems, it may well reduce cost as well as impact on the environment. This has been indicated through experiments performed by Xerox in Europe, in which used copiers were disassembled and the parts reused or recycled. Even using present copier designs and parts it has proved possible to reuse substantial amounts of parts and the profitability of the operation has been promising. In order to understand the business implications of this type of system operated on a large scale, a step change analysis could be undertaken. The application of reverse logistics and reverse factories represents a further step of systems development, beyond the development of new energy technologies, vehicles and machinery.

In the transformation program it makes sense to start to roll out the technologies in increasing volumes that already exist. At the same time, existing technologies could be adapted to new areas of application, and implemented on a large scale when adaptation is finished. In contrast, the development of new breakthrough technologies has to be planned for a longer term.

One early example of the time that it may take to retool and reorganize production facilities can be taken from the automotive industry. In 1927 Henry Ford had to close down his plants for 18 months in order to retool them for the new model, the Model A. In those days, production facilities were less specialized than they are at present. Today, production of a new product requires a large number of highly specialized tools for the pressing of metal, molding of plastics and other production steps at a large number of different companies in supply chains. Therefore, a large scale-up of production of energy-efficient vehicles and equipment will probably take several years, unless some production steps can be simplified and performed with less specialized tools. For this reason, as we will discuss below, it is important not only to understand which alternatives are best from an energy perspective, but we also need to understand the time and cost of developing new technologies and products, increasing production volumes and the issues related to market penetration of new technologies and products.

If unusually many companies were to need retooling for new products and technologies during a short period of time, the tool making would represent a serious bottleneck for the energy transformation program. In such a case, this bottleneck would need to be managed by the program, so that the most important projects get their tools first, and so that all tools involved in an important project got the same priority. This is another example of the complexity of large-scale change of complex energy systems, which require a certain amount of overall planning.

15.3.1   Systems and Stand-Alone Technologies

One of the aspects that make technology change complicated and risky both for companies and for their customers is that many technologies exist as parts of a system, where many things need to change in order for a certain technology to become successful. In cases where a technology or product can be changed without any dependencies to other products or technologies, we can see this as a "stand-alone technology" or a "stand-alone product." If a number of different technologies or other aspects of a system need to be changed, because of technological or systems interdependencies, we could term those "system-based technologies" or "system-based products." The transformation of areas where system-based technologies, and products, are used will be more complex, and, potentially, take longer to complete, than the transformation of areas with stand-alone technologies or products. In the case of system-based alternatives, the transformation process will probably need to be planned and managed in some way, unless systems can be broken up, or made less complex, in order to make decision making and change less complicated. This will be discussed in more detail in the last part of this book.

We need to look at some examples of each of these in order to understand the complexity of change in cases where system-based technologies are present.

15.3.2   Stand-Alone Technologies

A new heat pump in a house could be used as an example of a stand-alone technology. This heat pump could be installed regardless of any other installations in the house or at the neighbors'. It is run on the ubiquitously present energy source of electricity and nothing, except for the heat pump, needs to change. The owner of the house could decide to buy and install the heat pump without consulting with anyone else or waiting for any decisions from other people, or authorities.

Most situations where old equipment is replaced by new could be treated as installations or replacements of equipment based on stand-alone technologies. In most cases, it is only a matter of getting the funding for the investment in place, and one person, or in a company sometimes a team of decision makers, could decide to make the investment.

15.3.3   System-Based Technologies

In the case of system-based technologies, it suddenly becomes a little more complicated. A system of technologies works only if all the technologies that are included in or necessary for a system are in place and work. We also have a supply situation that needs to be taken into account. If an automotive company invests heavily in an engine technology that relies on a new biological fuel, this company must be able to count on the fact that the fuel in question will be available in large enough quantities over a period of several decades, in order to be able to sell the cars. In the same way, the customers who consider purchasing one of these cars are going to make the same assessment. If the supply of the fuel in some way seems at risk, the customer may decide to buy another car instead. On the part of the fuel producers, and distributors, they will eagerly watch the sales development of each type of car and each type of fuel, in order to understand which alternatives will be viable in the future. If sales figures for vehicles develop slowly, fuel producers and suppliers will hesitate to invest large sums of money in production facilities and distribution.

This is only one very simple example of a systemic dependency between technologies. There are many more. In order to make the discussion even more complicated, we need also to make a difference between stand-alone systems and integrated systems.

15.3.4   Stand-Alone Systems

A house is an example of a stand-alone system. This is a system that contains a number of different technologies, but the choice of technologies in one house is not dependent on the selection of technologies in other houses. Any person who builds a house can select the technologies and products for the house that he or she needs in order to build it and make it work. She does not need to consult with anyone else, except with the architect, the builder and the suppliers of the different parts of the house in order to build it. In case somebody wants to build an ecological house, of a type that has never been built before, nobody can stop them as long as it is built according to existing regulations.

A stand-alone system is not dependent on the adaptation of other technologies outside of the system. In the case of a house, it needs to be connected to the electricity grid, sewage systems and other systems, but these systems do not need to be altered depending on the type of house that is going to be built on the site. In principle, the owner of the house needs to adapt the interfaces between the house and these systems to the standards of each of the systems.

The same is true for a power plant that is based on cogeneration, photovoltaic or wind energy. When a new power plant of any of these types is built, it is expected to deliver electricity to the grid in the same way that the other existing plants do. We do not need to ask customers of the electricity company to buy new white goods in order to make it compatible with the new power plant. This makes it relatively easy to make decisions about the investment in a stand-alone system, compared to decisions to invest in a technology that will be part of an integrated system. In the case of power plants that are based on natural gas or biofuels, these are parts of truly integrated systems. In order for power plants of these types to become viable, the fuels that are going to be used have to be available for the long term in large enough volumes and at low enough prices.

15.3.5   Integrated Systems

Some of the energy systems that need replacement are integrated systems. In such a system, a single supplier to a part of the system cannot decide to change important aspects of this part, without the active support of other suppliers to the system. Obviously, systems consisting of engines and fuels belong to this category of system. A supplier of one technology is often cautious to invest in new technologies before information about the plans of other participants in the system is available. In a situation in which a large number of opportunities exist at the same time, it becomes very difficult to make decisions about investments in the development of new technologies. In such cases, development could be slowed down for years, with companies waiting for sources of uncertainty to disappear. Still, if companies do nothing or almost nothing, very little learning takes place and large-scale action is delayed.

This is one of the situations in which we need an overall plan in order to avoid delays and large financial losses, because companies invest in technologies that fail to survive because of unforeseen systemic interdependencies or miscalculations. I do not suggest that governments need to turn the market economy in such sectors into planned economies, only that a certain amount of security concerning the future is needed in this type of situation. In particular, in order to develop new engine technologies and expand the production of vehicles that use these technologies, automobile companies need to know that there will be a network of filling stations or charging stations (in the case of electric cars) across the country or globally. In order to build this network fuel companies or electricity companies need to enjoy a certain amount of certainty about the future actions of vehicle companies and legislators. In order to make investments, investors need to know that there will be enough customers that demand this type of fuel. This means, probably, that several automobile and truck companies need to invest in the same engine technologies and that several fuel companies or other suppliers of energy to the system will need to invest in production and distribution of the necessary fuels. All in all, for this to happen in each country, governments will need to decide about the general direction of their fuel policies and on an international level there may be a need for agreements between countries about which fuels each of them will prefer to promote.

The hope that the market will be able to settle a competition between close to a dozen different future alternatives, without undue delay and horrendous loss of money, will most probably prove to be futile.

15.3.6   Changing Supply Chain Systems

In another type of integrated system, the situation is different, but in this case as well, companies may need guidance from overall policy makers, before they decide to make major changes themselves. This is in the case of replacing existing global and national supply chains by an increased amount of local and regional supplies. As we have seen above, many industries have moved toward global supply chains and production networks. In many industries specialization has been driven very far and production has been centralized to a few units globally, which require extensive transportation networks in order to deliver supplies into plants and to move products to market. In other industries, production is still regional to some extent, but it has usually been concentrated in order to achieve economies of scale in production, at the expense of increased transportation. Production and transportation networks are examples of complex integrated systems of highly specialized producers, suppliers and production networks, where it will be difficult for one player in the system to make decisions without information about the general direction of the development of the system as a whole.

In many cases improved energy efficiency will require investments and, perhaps, cost increases, which will be difficult to accept in the short run, since such decisions will affect the long-term competitiveness of companies. At least managers will need reliable forecasts of future energy prices, and forecasts of the energy supply, but in many cases this will not be sufficient to make major investment decisions. This means that no CEO in his right mind could make a decision to move production to a more expensive country or change suppliers in a similar way, simply because of environmental consciousness. He will need a long-term business case in order to do this. This is not because managers are not aware of the perils of decreasing oil supply or climate change. Managers are simply appointed by the board with the remit to make a profit. This means that companies will focus on financially justifiable investments. Companies and their managers could spend small sums on charity, but they could not, within the existing system, make large investments based on other goals than making money. A long-term business case could be based on an overall plan for the energy transition, which has been described above, which needs to be firmly grounded in business fundamentals. Such a plan could, in the best of worlds, indicate the level of change that will be needed for the longer term, in order for society to adapt to new conditions.

15.3.7 The Economic Life of Investments

Another aspect that, in this case, may complicate the transition is that, in order to run the process toward energy efficiency as rapidly as we need to, we may be forced to replace existing technologies before they have finished their economic or technical lives, or upgrade to a technology that is more expensive to purchase or operate, at least at the launch of the new technology in the market. To many managers this is an unthinkable solution and it cannot be expected from companies that managers will voluntarily invest shareholders' money in the implementation of technologies at a point in time at which it does not make financial sense. While technically it is the company that emits pollutants that are polluting the environment, from an economic perspective it is the workings of the market and the demand from customers that give rise to certain production systems and business practices. It may be a simple solution to make the polluting companies pay for the investments in clean technologies, but it could be argued that we have all caused pollution through our consumption habits and, from an economic perspective, companies have only responded to market demand by producing at the lowest possible cost, using existing technologies.

Now, many people and some governments expect companies, such as the ones in the Swedish pulp and paper industries, to invest in new processes that cause less pollution, and that use less energy. It is unlikely that this will happen rapidly and on a large scale without regulatory measures, subsidies or at least an overall strategy for society, which indicates that all participants in the economy will, in some way, have to share the burden for the energy transformation. In most cases, once an investment has been made, the owner of the equipment tries to use it for as long as possible without replacing it, and when a technology is about to be replaced, it is usually replaced by the most cost-effective alternative in the market, which may or may not be the most energy-efficient. Many times installations can be used for decades and companies, individuals and society as a whole prefer to invest in new facilities, increased capacity and other new investments, rather than pay for the maintenance and upgrade of old installations and equipment. This makes sense, because using existing equipment for as long as possible increases our wealth, as long as we maintain and upgrade the existing equipment in order to keep pace with competitors. It has been mentioned previously that the median lifetime of a car and light truck is around 17 years and it takes another 10 years to replace the median heavy truck. In the transformation program, it may be necessary to increase the rate of replacement of both vehicles and other investments, but that would put a heavy strain on the economy.

In the case of some technical installations, such as power plants and energy distribution systems, they have economic lifetimes of several decades. The same is true for buildings and production facilities. In all these cases, owners and operators are reluctant to replace existing facilities. In many cases, it will be possible to refurbish existing plants, buildings and other facilities with sustainable technology. In other cases, however, the whole complex may need replacement, which will be a much more difficult decision. In these cases, as well, we must not only compare the technical advantages or disadvantages of different technologies. We also need to compare the size of investments in relation to the environmental benefits, reduced energy consumption or clean energy production potential, when we decide which technologies to use.

15.4　INCENTIVE CHANGE AND CHANGE IN CONSUMER BEHAVIOR

In companies there are many forms of incentives in place. Some are explicit and some are not. Many incentives are founded on values. People are sometimes promoted because they look busy and give the impression of being "on top of things." This results in an implicit incentive to look busy and talk in a self-confident way. Managers who are responsible for a number of offices may feel that they need to be seen in remote offices at certain intervals, so that people feel that their office, too, is important to the boss. This may be done in order to avoid the spreading of rumor, to the effect that a certain manager only cares about the people in her own office, which could put a future promotion at peril.

In order to change these things on a large scale in a company or across society, incentives to save energy need to be put in place. Some incentives can be decided internally within a company and promoted by its employees, but the most important and immediate incentives are provided by customers.

As individual values in society change, companies will start to feel pressure from consumers to change their ways of doing business. This is part of the ongoing processes in a market economy and is important for economic development. As customers in the early years of the twentieth century demanded more cars, the demand for horses, carriages and other items related to old ways of transportation experienced a shrinking market. If consumers demand increased energy efficiency in their lives, some businesses will grow and others will shrink. Companies on the supply side will need to position themselves and take strategic measures to benefit from the development, rather than suffer from it.

The most important source of incentives in a market is provided by customers. Customers may in the future demand more products that have been produced in energy-efficient processes, because they want to contribute to conserving energy, or because, in many cases, products that have been produced using energy in an efficient way may be less expensive than competing alternatives. This advantage of energy-efficient products and services will increase as energy prices increase or as the supply of oil decreases.

Simply by knowing that energy conservation is becoming a priority, customers may increasingly turn away from products that they know to be energy-consuming. Examples of this could be gas-guzzling cars, imported fruit and vegetables and red meat. Also in other product categories than food, consumers may turn away from imported products and prefer locally or nationally produced products, even if they sometimes cost more. As this process develops, companies may want to inform customers about the products and services that are more energy-efficient. In many situations, systems for marking products in stores with their energy content from production and transportation could help in achieving this. Such marking could either be implemented by stores on a voluntary basis, or it could be part of a transformation program, which could also include information activities to consumers, that could provide guidelines for how consumers should use the information provided on labels.

As customers become better informed, they can decide to increase purchases of goods that demand less energy and decrease consumption of goods that leave a big energy footprint. When this happens, the signaling through the system will be rapid and the effects of these changes in demand will be felt not only by the company that produces the consumer product but also through changed purchasing habits by the consumer goods company, by its suppliers of machinery, raw materials and components. In due course, companies that experience changes in demand because of changes in priorities among its customers will, in turn, develop systems for evaluating suppliers not only based on price and the quality of their products, but also based on the energy efficiency of the production and logistics processes of the suppliers. Demands from quality systems on systematic reduction of energy use in different areas will increase.

Even without customer pressure, many companies, as mentioned above, will be able to improve energy efficiency in many respects. The efforts to do this will, probably, be weak in the absence of strong incentives. Increasing energy prices will in many cases not be sufficient as incentives as long as there are market opportunities or other savings opportunities that could provide a higher return on investment. Some incentives and guidelines may be provided internally by corporate management.

Companies could include activities to improve energy efficiency in the evaluation of managers and, for example, put in place guidelines and procedures that make it possible for managers to be favorably evaluated by employees in remote offices, without their frequent presence there. Such procedures may include information internally that energy conservation has become a corporate priority, and that managers, perhaps, are expected to travel less. Admittedly, corporate travel for most companies represents a minor item in terms of energy consumption, compared to production and distribution, but it is a highly visible item. Corporate decisions to restrict travel affect the behavior of people. This may influence the behavior of employees in their private lives more than decisions to install a heat pump for heating the office or a plant. Big-screen television sets and video conference equipment could be installed in offices and plants so that managers can be seen and can talk to employees without being present. Companies could also reduce travel budgets and reward people who spend less time traveling.

Box 15.3 The Seemingly Obvious Improvements May Not Be the Most Important from an Energy Perspective

In this book a large number of different opportunities to conserve energy and reduce pollution are described. We need to develop new technologies, products, services and business practices, in a number of areas, in order to achieve the savings that we as a society need to achieve. Probably, it will not be enough to develop new technologies, we will also need to reduce consumption or change our lifestyles or work practices in different ways in order to achieve the rate of savings that we need. One of the main arguments is that we need to prioritize different actions at the level of society, in order to identify the ones that will provide the most benefits at the lowest cost.

This type of prioritization is already going on in many large companies, but the knowledge that has been developed in one company has often not been shared between companies and it has also not spread to smaller companies or to society as a whole. One example of the results of such an analysis could be taken from a global company that has found that the global business travel of its employees has a relatively low impact on the environment, compared to other activities. This company has also found that a person does not save energy by reading a newspaper on the Internet, compared to subscribing to a physical, paper-based alternative. The energy that is consumed by the computer while reading the paper is, at present, as high as the energy consumption in the production and distribution of the newspaper in the paper format. As new technology will be developed and made available to producers and consumers, the energy consumption of both these processes will change.

It is important that the savings opportunities of different measures are analyzed, but there may also exist invisible psychological links between different types of behavior. Private driving and leisure travel consume larger amounts of energy than business travel. We will need to reduce petroleum consumption and it is better if we start to reduce this consumption ahead of dramatic price increases. Petroleum products are primarily used for transportation, fertilizers and pesticides and for plastics production. Transportation breaks down into transportation of goods and of people. Plastics break down into high-value and low-value products and many times it will require more energy to replace the plastic of high-value plastics parts in an office printer by a similar part made of metal.

It is important that we start to draw the different decision trees and debate which savings opportunities will cause the least pain to people and economic systems. If we want people to change their behavior we need to understand the psychological relationship between different types of behavior. We may have to start to change behavior in small but visible ways, in order to achieve large-scale behavior change when people have become used to the thought that change is necessary and possible.

With increased incentives to conserve energy, companies will also have incentives to invest in new and energy-efficient technology. As will be explained in more detail in Part III, technology change will be a long-term prerequisite for a sustainable systems change.

15.4.1 Government Incentives

In order to drive energy transformation, governments will, most probably, have to make a plan for the transition and include a number of different financial incentives and tools, directed at businesses or individuals in the package. Some examples of incentives have been mentioned earlier in the chapter about financial tools.

These tools could be used in many ways in order to provide incentives for both businesses and households to change consumption habits and invest in new technology. In principle incentives could be used for all situations. However, as in all the cases in this book, incentives from the government and from corporate management, toward employees, need to be applied in a selective and controlled manner, giving priority to incentives in areas where they are most important.

15.5 SYSTEMIC CHANGE

The transition toward energy efficiency will be gradual. Step by step we will need to adopt practices and technologies that will change a small aspect of life and of society in the desired direction. Over time we will find that the society of the future, in many ways, will be completely different from the society of today. The change will not be limited to technology changes or to changes in consumer behavior. As change progresses, we will experience completely new sets of values and behaviors among leaders and among people in general. This will happen because many things that did not make sense in the past will make sense in the future. It will make perfect sense to care about the environment and reduce energy consumption, but it will not make as much sense to drive a gas-guzzling car. These are only simple examples that, like the tip of the iceberg, will be signs of more unexpected changes underneath.

Some aspects of society may become reminiscent of days past, when business was more local and many products were produced in the area and sold in the local market. Some prepackaged products may disappear and be replaced by local produce sold by weight and packaged at the point of purchase, in modern systems, for this purpose. Some industrial products may disappear altogether, because they need too much energy for production or distribution, or because they do not fit in the lifestyles of the future.

However, this scenario of large-scale lifestyle change and supply chain transformation may be contrasted by another alternative that could be more focused on technology change and the wide implementation of new sustainable technologies. We may be able to succeed by going down any of the routes, or a combination of both. In any case, the two alternatives, if seen as alternatives, may give rise to two very different societies of the future. If the alternatives are used in combination, different combinations could also render substantially different societies of the future. The choice between different principles toward a sustainable future, choosing between alternative systems, could be a new arena for political debate in the near future. As we start to discover and understand the different alternative systems, which we may want to create, different parties, or factions within parties, may argue in favor of different alternatives, and we may see a new interest in politics arise from these new opportunities. We will look into some of the alternatives below.

During the change, governments may have to intervene to a small or large extent in the workings of the market. The forms of government intervention may be decided through political debate, but the debate, in order to be effective, needs a firm basis in the analysis of alternatives. After a period of some government intervention in markets and planning of change, this intervention may once more disappear, similar to when any other project is finished and the members of the project organization go back to their regular tasks.

The level and direction of change are difficult to foresee at this point. We have far too little information about the speed of change that will be needed and the opportunities at hand in order to argue that some actions will be necessary and others will not be. Some people may at this point argue that we can maintain a global economy and global supply chains if only we improve the fuel efficiency of vehicles, airplanes and other means of transportation. Others may argue that we will need to completely change the economy into a set of local self-sufficient economies. From an economic perspective, we currently have no widely shared basis that we can use to settle such an argument.

Box 15.4 Stable State Economics

A form of economic system that contains a number of mechanisms that are completely different from our present system has been put forward by a number of visionary economists, who argue that our present system will outgrow (or has already outgrown) the finite resource base of the earth and needs to be replaced by a system that does not demand constant economic growth. The proponents of such a system often quote the economist Kenneth Boulding: "A person who believes in infinite growth in a finite system must be either a madman or an economist."

One of the most prominent proponents of alternative economic systems, based on other values than constant growth, is the economist Herman Daly. In his book *Beyond Growth* he outlines the idea of a system that he calls "stable state economics." While this is an interesting and probably a necessary idea, it does not seem to be fully developed and the practical road of transition from the present system to a stable state system is less than clear. Like many of the energy-efficient technologies that we need in the future, which need to be developed before we can use them, ideas such as stable state economics need to become better defined down to the minute details or tested on a small or medium scale in order to make their actual strengths and weaknesses visible. At present and probably for the foreseeable future the resources within the global economies, which are necessary in order to succeed with the transformation, are tightly linked to the existing economic system and this system will remain necessary for large-scale energy transformation. One important example in point is the Ecomagination project run by GE. In this project the company since 2004 has been investing 700 million dollars per year, a figure that is gradually increasing to 1.5 billion dollars by 2010. This type of investment in the development and distribution of new technology that will be necessary for the energy transformation could not easily be envisioned within a stable state economic system. Even though 1.5 billion dollars annually is a significant amount of money, this figure needs to be increased many times in order to make possible the investments necessary for the energy transformation.

Complementary currency systems and other financial innovations will increase in importance in order to strengthen local economies and reduce the pressure on national currencies and the global economic system, but they will remain fringe phenomena unless a severe crisis in the global economic system or a dramatic shift in sentiment among citizens give rise to a completely new situation.

Regardless of which level of change we believe in, we will experience a certain level of systemic change, and changes in values, production systems and technologies. Society, as we experience it today, will not be the same in two decades. We will now look into two of the alternative routes that exist in principle.

15.6   TWO PRIMARY ALTERNATIVES

Energy transformation could be based either on the implementation of new sustainable technologies, or on the change of behavior among consumers and companies. One alternative will not exclude the other, and it is more a matter of which alternative is going to be at the fore and drive change and on which aspects governments and steering committees for the overall change program will focus their efforts. Should overall management of the program focus on the implementation of financial incentives for the development and implementation of new technologies? Maybe the government should even make direct investments in technology development, as in the case of military and space technologies of the past. Or should they focus on providing consumers and companies with information and incentives to change their purchasing behaviors and other behaviors in the market, and allow the signals provided by customers to guide companies to develop the preferred new technologies?

These alternatives could be debated at length, and it is important to recognize that each of these alternatives gives rise to different change processes, and probably also to different results, in terms of the societies that emerge through the change. In the near future the debate may even give rise to different parties that advocate one or other of the alternatives.

15.7   TECHNOLOGY-BASED CHANGE

The development of Western society through the past decades has been driven by technology change. To a large extent, the development of new technologies, such as nuclear power, the computer, space and aviation technologies, has been financed by government. This development has spawned a large number of new industries and companies.[1] If it were not for this research and development, we would have had neither the current type of society nor the standard of living that we are now enjoying.

As a society, we have a very good understanding of the consequences of technology development, and also of the financing of this type of effort. Therefore, it is probable that many people would be in favor of this type of development. It has led to economic growth and prosperity in the past, and it is probable that we would be able to make it work for us once more.

When we discuss investments in technology development and implementation, within the program, we need to distinguish between large-scale and small-scale technologies. The reason for this is that there is a difference in the financial and administration tools that are necessary for the two types of technologies. Large-scale technologies are represented by those that need large-scale implementation projects in order to be implemented. Examples of large-scale technologies are power and heat generation technologies for large power plants, and the development and implementation of new engine technologies and renewable fuels, which, in many cases, could only succeed if these projects are run on a large scale, with the aim of creating a large market for any particular technology. Small-scale technologies are those that could be implemented by single households or by neighborhoods, such as small-scale wind and solar power generators, heat pumps and energy-efficient light bulbs and other technologies. Many products, based on small-scale technologies, have a low level of complexity and could be developed and marketed by relatively small companies. The level of complexity of the change programs that will be necessary in order to get large numbers of households, individuals and small companies to change may yet be high. In order to achieve high volumes and rapid market penetration of such technologies, the commercialization phase may need to be run as a large-scale project of capacity expansion.

15.7.1 Change Based on Large-Scale Technologies

Large-scale technologies require large projects, with a large share in the projects taken by large companies. The large-scale assignments for technology development, production and implementation of large-scale technologies would be similar in size and structure to the projects run within the space program, in which large industry incumbents in many industries have participated.

Large-scale technology projects for power generation systems, and fuels and engines, would solve emission problems for a large number of consumers or companies in one go, on a large scale. We may also think of large projects, involving the construction of a number of power plants, using sustainable technology, which would be built in a number of urban areas at the same time. Each project may involve investments of hundreds of millions of dollars, and this would reduce emissions for the power and heat generation for a large number of households at the same time. We could also think of a project involving the implementation of a new public transportation system, which could be implemented in several regions at a time and solve the transportation needs of a large number of people.

This type of project is very large, but it requires a completely different type of administration, and project management, compared to small-scale technology projects. It could be argued that they also contain a different type of complexity, since they mainly involve technology companies and the management of technology development and production. The projects would represent a number of large-scale construction projects in a number of different geographical locations. These projects could also involve the manufacture of production resources for the necessary renewable fuels.

The reason for the government to be involved in the management and financing of this type of project would be that the size of the investment in many of these cases, and the risk involved for any company that would try to undertake it under pure market conditions, could in many cases (as is probably the case for renewable fuels and engine technologies) be too large. In some cases, political decisions about the content and structure of future production and distribution systems may be sufficient in order for large companies to invest. In other cases direct investment in technology projects may be necessary in order to drive development forward. It could also be the case that the payback time for investments, as in the development of general purpose technologies, may be too long, or it may involve the replacement of resources that have not yet finished their technical or economical lives. In any case, as has already been argued, it would be difficult for many companies to voluntarily drive this type of project at the high speed that will probably be needed in order to rapidly enough reduce energy consumption.

In the case of engine technologies and fuels, we are looking at projects that involve technology development and the creation of systems for the mass production of vehicles that will be using the new technologies. This type of project may also need to establish a new network of filling stations or charging stations for the vehicles, and production and distribution facilities for fuels. The need for large production facilities for vehicles and fuels and a large-scale geographical network of filling stations adds aspects to this type of project that are not present in the case of large-scale power plant projects.

These different aspects of complexity of projects, and the challenges that need to be mastered in each case, influence the risk, investment and cost of projects. They also influence the time it will take in order to realize the potential energy savings of each project. As we will see in Part III, each alternative technology, and the projects that will be based on each technology, will need to be evaluated on its own merits.

15.7.2   Change Based on Small-Scale Technologies

In the case of small-scale technologies, technology development is not as complex as in the case of large-scale technologies. Small solar or wind power systems, or heat pumps and other small-scale solutions, can often be developed and marketed by relatively small companies that sometimes operate in markets that are geographically limited. The opportunities, in this case, to develop and rapidly grow breakthrough technologies in markets with small companies will probably be limited, but large companies could, of course, also develop technologies in these areas. The existence of a number of different solutions, and the ensuing competition between the companies that will be involved in the development, production and distribution of each of the technologies, would create a more dynamic market, with better prospects for technology development.

The number of customers that could demand this type of technology is large. The critical aspect of this type of project would be to rapidly increase the market penetration of energy-efficient technologies. In many cases, small-scale technologies could be rolled out in parallel with large-scale technologies for purposes such as power generation and heating. Small-scale technologies would be relevant, at least, for customers who live in areas without large-scale solutions, or if the large-scale solutions prove to be too expensive to implement or use for some segments of the market.

Box 15.5 Subsidies for the Implementation of Small-Scale Energy Technologies

Sweden used to be heavily reliant on oil for its substantial needs of heating in the winter. During the 1980s many households switched over to electricity-based heating systems. Due to the increasing demand for electricity, and the cost of expanding the resources for power generation, the Swedish government has subsidized the installation of small-scale renewable energy, and energy conservation, solutions for individual households, small apartment buildings, and small companies. This has included subsidies for the installation of heat pumps, boilers fueled by wood pellets, and other sustainable energy technologies and investments in additional insulation, triple glazing windows and other energy conservation technologies.

These subsidies of small-scale systems have been complemented by the construction of large-scale heating and cogeneration plants and distribution systems for heat within the districts of many cities and towns.

The government could get involved in this type of project in order to ascertain a rapid enough construction of production resources, and a rapid enough penetration of these technologies in the relevant segments.

While the advantage of these technologies will be that they do not require huge resources for the development of the technologies or products, the disadvantage of small-scale solutions may be that the smaller companies that get involved in the development and commercialization of these technologies may not possess the financial resources that are necessary in order to rapidly build production capacity and market demand. The existence of an installed base of old and less energy-efficient technologies, which may still be possible to use, although at an increasing cost as energy prices increase, will also act as a drag on sales, unless incentives are launched for consumers or companies to invest in these technologies. One very important aspect will be the large-scale marketing of this type of technology, which needs to be planned in parallel with the scale-up of production resources.

15.8   BEHAVIOR-BASED CHANGE

Large-scale and controlled behavior-based change is often debated in the literature on peak oil and climate change. We need to cut down on traveling, drive less, change our eating habits or production and distribution systems for food, so that we eat more food that is produced locally using less energy, and live closer to where we work. As a society, we have less experience of controlled large-scale change of habits and lifestyles, over a short period of time, than we have of technology-based change programs. We also have experience from the economic effects of technology change, and the financial and other alternatives that are at our disposal to drive transformation. This aspect is important, because our current economic system is dependent on economic growth, and it may be difficult to sustain economic growth if we dramatically change our lifestyles through a dramatic reduction of consumption.

This is not to say, that behavior change should not be attempted, only that the tool kit for driving behavior change will need to include a set of economic and financial tools that could be implemented in the economy, in order to mitigate the economic consequences of this type of change. More than any technology-driven change effort, the training and information program directed at consumers and employees in companies, which must form a core part of a behavior-based change program, needs to be built on a state-of-the-art understanding of economics, marketing psychology and change management.

These aspects of a change program that would involve large elements of behavior change have to a large extent been overlooked by the proponents of such change. The ideological aspects of behavior change have been more strongly emphasized, such as the aspect that we have no right to exploit our planet in the way that we do in the Western world. While this may be true, we have to make sure that we do not destroy the life-sustaining virtuous cycles that we have managed to create that do now supply food and other necessary resources to the more than 6 billion inhabitants of our planet. Instead of putting these virtuous cycles at peril, we need to improve our understanding of them, and gradually make the economic and production systems of the world more robust and sustainable.

15.8.1   Change Based on Individuals Who Change Their Behavior

There are many opportunities to save energy, through the implementation of large-scale programs to change behavior. The change program could involve the information to households that they need to change part of their lifestyles, such as cut down on the consumption of those foods that require much energy for production or distribution, change their driving habits or reduce the number of miles traveled for vacation. These types of demands on people are normally very difficult to get across, without the risk of alienating large parts of the public. For this reason, the planning of information campaigns containing this type of message would be very important, and it is also important to select the primary, secondary and tertiary audiences that would be targeted by different messages.

In any marketing campaign, it is possible to identify the early adopters of an idea, the early majority, which is the first high-growth segment reached by the rollout. After the early majority comes the late majority, and then the so-called laggards who adopt a new idea, product or service late in the adoption cycle. In this case, as the goal will be that the idea of energy conservation should be adopted by a certain share of the population, it is important to understand how different groups in society perceive the idea, and how they can be persuaded to adopt more energy-conscious lifestyles.

Experiences from the adoption of earlier products and services, with some features that are similar to that of energy efficiency, should be taken into account in the planning of information campaigns. This analysis of previous campaigns should include both successful launches and less successful ones.

One successful campaign, which has served as the role model of many campaigns since, was the campaign to launch the light beer Miller Lite. In order to counter the expected resistance toward a light beer, and to market it to, mostly male, drinkers of regular beer, commercials showed popular athletes drinking Miller Lite, with the goal of staying fit. The initiative by former Vice President Al Gore to step forward and serve as a role model in the energy transformation program could be seen as a step in the same direction, but this step needs to be complemented by a well-planned information campaign, which could inform citizens across age groups, incomes, and social strata of how they could contribute to increased energy efficiency.

The content of such a campaign needs to be developed by a team that includes categories of experts, such as marketing experts, economists and representatives from retailing. Throughout the energy transformation program, it will be important to build consensus around measures and projects. We will need to constantly make trade-offs between rapid energy savings and the need to manage change in the economy in a structured and predictable way.

15.8.2 Work Life Change

In addition to the opportunity for companies to invest in new technology, there is also the possibility of changing the behavior and routines of corporate life. The opportunities of reducing energy consumption, through better planning of travel, the selection of suppliers based on the energy content of supplies, in addition to using the criteria of cost, delivery time and quality in the evaluation of suppliers have already been mentioned. In many cases, these behavior changes also represent cost savings, which will become more significant as energy prices increase.

In the same way as in the previous case of individual behavior change, it would be important to understand the market before a campaign is launched. Who are the people in companies, who have the largest influence on energy consumption, and how can they be persuaded to reduce consumption themselves, or persuade others to do this? Naturally, managers play an important part in most company decisions, but there may also be other people involved, who have formal or informal authority in this regard. Managers of different departments and at different levels may have differing levels of opportunity to influence consumption.

It needs to be repeated that while large-scale technology change is relatively well understood, both from the point of view of managing this type of change, and from the point of view of understanding the consequences and the risks and difficulties of this type of project, a project involving large-scale behavior change on the part of individuals or employees in companies contains many more unknown factors.

These may be aspects related to how the message should be formulated, and who the sender of such a message should be, or aspects related to the gradual or sudden adoption of the ideas and the short- and long-term consequences of this adoption. The large-scale change of individual consumer behavior will involve more unknown factors than the large-scale change of company behavior, and the communication aspects will differ between the cases.

It is probable that all of the above four basic types of change projects (small- and large-scale technology change and change of individual and corporate behavior) need to be used by the energy transformation program. The exact timing of each type of action and the extent to which each type of action should be applied at any one moment need to be well analyzed, planned and discussed. Many things can be achieved in only a short period of time. The projects with the largest improvement potential, showing the lowest cost per unit of energy saving, should be prioritized, and the more costly or problematic changes could be made in due course.

15.9   ECOLOGY AND ECONOMY CHANGE

The energy transformation program is a long-term process to transform our heating, electricity, fuel, production and distribution and other technological systems and create a sustainable society, from the energy perspective. We look forward to decades of technology development and investment schemes and transformation sector by sector. Some changes can be implemented at short notice and at little cost and investment need, but many changes require a number of steps until we are be ready.

Experts in different ecological disciplines argue that we cannot allow CO_2 emissions and global warming to get worse, without harming the mechanisms that could help natural systems regain their balance.

According to these experts, we only have a few years to reverse the processes, or many of the changes may become irreversible. This requires dramatic and focused change. The awareness of the impending problems may be enhanced through the increase in energy prices that is created by decreases in oil production.

As important a goal as that of sustaining ecological systems, but one that has been less widely recognized up until now, is the one of sustaining economic systems. In a situation where we face the risk of declining energy supply in the future, we need to secure existing production and distribution systems, as well as maintain the virtuous cycles of the global economy, which provide the bases of life for more than 6 billion inhabitants on this planet. Major changes, therefore, need not only to be rapid to save the environment but also well planned and managed to maintain virtuous economic cycles.

The overall goal of the transformation program needs to be that of creating a sustainable society, with sustainable and renewable energy systems for the future. If we achieve this, rapidly enough, all that remains is to hope that the climate will return to normal, and that the changes that we have caused will do as little damage as possible to ecological systems. At the same time we need to hope that we will be able to maintain the most important aspects of our economic system, namely, the ability to cater to the needs of a huge global population.

CHAPTER 16

Change in Different Sectors of the Economy

In order to understand the complexity of the change program, or at least the complexity of defining the strategy for the energy transition, we need to recognize another fundamental difference, in terms of the ability to change, between energy consuming units.

16.1 LARGE COMPANIES

The part of the economy that is always easiest for government to manage, monitor or inform is always the one that is represented by large organizations. These organizations can be addressed and informed through communication with their top management teams, and formal organization structures and mechanisms can be used in order to spread information inside each company. This way, through this one point of contact, information and plans could be communicated to thousands of individuals.

This makes any project aimed at large companies both comparatively easy to manage and cost-effective. In addition to this, large companies have the financial resources themselves to make operational improvements internally that could pay off in the form of long-term cost savings and improved competitive power. They will also have the financial strength to develop some energy-efficient products and services. During the Second World War, for instance, Boeing invested large sums of its own money in the development of some of its bombers for the air force.[1] As in the case of The Ford Motor Company in the war economy and Boeing and other aerospace companies in the space program, large companies could also be assigned important tasks in a transformation program, as producers, or developers and producers of new technologies and products.

All in all, large companies represent both an area for the transformation program, where quick and not too expensive wins could be achieved, and important resources that can be utilized throughout the program for various tasks and efforts.

16.2   SMEs

With Small and Medium Sized Enterprises (SMEs) it is a totally different picture. Experiences show that the bulk of SMEs do not themselves have the resources to manage a transformation project and many times the savings potential in financial terms is so small that it does not justify a transformation project with substantial amounts of dedicated resources. Yet, SMEs use a substantial share of the total amount of energy that is consumed by the industry of any country. Only to a small degree, the experiences made in the energy transformation of large companies can, for different reasons, be transferred to this category.

The SMEs, by definition, employ less than 250 employees each and this community is very diverse. In terms of energy use, SMEs range from service companies that use energy mainly for lighting and heating, and the transportation of employees to the office and back home every day, to casting companies and other heavy industrial firms that run energy intensive production processes. There are also SMEs that indirectly consume substantial amounts of energy in transportation and distribution systems run by other companies, of which these SMEs are customers. The purchasing patterns, production and logistics systems run by individual companies could be transformed partly by the companies that buy transportation services, as they demand higher levels of energy efficiency and lower levels of pollution from their suppliers.

In order to address SMEs with information about energy transformation opportunities, the whole category of SMEs needs to be broken down into subcategories. Energy transformation opportunities need to be analyzed for each category, and transformation projects need to be run for whole networks of companies with similar business and production processes. This needs to be done in projects with the aim of developing methods and tools for transformation that deliver results in a cost-effective way. In these pilot projects, a resource of transformation consultants will have to be trained that could later train the next "generation" of trainers in the tools and methods as projects expand.

Most probably, a combination of medium- to large-scale training seminars and networking activities and on-site training and support at participating SMEs need to be provided by a program in order to be effective.

Part III

The Program Structure

CHAPTER 17

The Transformation Program

The energy systems transformation cannot be achieved without a certain amount of overall planning. The most important reason for this is the speed with which we probably need to transform energy systems, in order to avoid climate disaster and oil shortfalls. Most probably, we do not have the time to wait for the market mechanisms to slowly deliver results. At some point in the near future, we will discover that we need to speed up the process, and run the program as a planned transformation effort. At this point, we will need to rapidly get our act together and make a number of decisions related to the strategy and planning of the program.

Box 17.1 Planning Based on Constraints – Limited Resources

Any plan needs to be based on a number of constraints. No project has an unlimited budget, or other unlimited resources. The pressing global goals of the reduction of CO_2, and the warnings of alarmists, call for a plan, and when we make this plan, we need to recognize the constraints that are imposed by the limitations on resources for energy production.

In the energy transformation program, the two most obvious constraints are the constraints of time and money. We need to achieve rapid results. The magnitude of the change means that we will have to use large amounts of money, some of which will have to be invested long term, in projects with uncertain prospects for financial payback. This means that we need to prioritize investments that can deliver renewable energy, and possibly start to pay back on the investments, in only a few years' time, but we may also have to prioritize less certain investments. We also need to prioritize by the size of the investment and the volume of investment per unit of energy saving that will be delivered by the solution at different time intervals. In most cases, I have been unable to gather reliable and comparable data on this while writing this book. This has to be one of the tasks of the strategy development work in the program.

Other constraints that we need to acknowledge are the limited access to raw materials for energy production, and land for the growing of energy crops, or the construction of photovoltaic power plants, and wind turbines. For each of the solutions for renewable energy that we are looking at, we need to analyze how much raw material, or land, that we, at different points in time, will be able to set aside for the production of this type of energy. Needless to say, we also need to be careful not to double-count the use of any resource, in order to avoid arriving at unrealistic estimates for the production potential of any of the alternative solutions.

The complexity of the transformation is one more key aspect. The very large spectrum of different opportunities for change available to us is another. In order to achieve rapid and reliable results, we need to focus on a number of primary change goals that need to be defined in terms of areas for change, technologies and market penetration. These prioritized change goals need to be pursued in a planned and structured way, from the current level of technology development and market penetration, to the successful development of the necessary products and services, and the expansion of production capacity, to the marketing effort that will result in the large-scale implementation of these solutions.

As already indicated, the complexity of energy transformation arises from a number of different dimensions of the issue and the possible measures. There is no single solution that could solve the entire problem in a few years. The different aspects range from technology change and behavior change to the choice between large-scale and small-scale technologies, and the problem of managing a number of efforts toward goals that need to be both short- and long-term. There is also the problem related to the interrelationships between technologies. If we use less oil, we need more biofuels, electricity or other more radically new technology alternatives. We need to be able to handle the development of some, while we reduce our dependence on others. The issues are to some extent technical, but they are also related to economics, business and project management.

Many people believe that the solution will be to dramatically increase the volume of nuclear power, so that we can also run cars, trucks and other means of transportation on electricity. The problem is not only that – in many countries, there is a very strong public opinion against nuclear power. There is also the problem that it may take a decade from the start of the project until the first new nuclear power plant will be finished. If we start planning such plants today, it will take many years for us to install the energy production capacity that will be needed in order to produce enough electricity to fuel global vehicle fleets. Some experts estimate that this would require 500 new nuclear power plants only in the United States.[1] In addition to this the supply of uranium to fuel these plants is also limited, but using the most modern types of breeder reactors the supply will probably last for more than a century. A similar situation exists for wind power plants. If we were to turn entirely to renewable energy sources in a few decades, we would need to invest not only in a large number of new wind turbines and photovoltaic power generators, but we would also rapidly have to build production capacity for wind power turbines, which is currently far too small to cater to future needs.

Following another line of reasoning, some people argue that the solution to our future energy needs will be unconventional oil, produced from oil sand or oil shale. This does not seem to be a large-scale solution. The Canadian government expects that by 2025, the production of this country, with the second largest endowment of oil sand in the world, will be 3 million barrels per day, which is 4 percent of the current global oil production.[2] Other people argue that large deposits of methane, bound under the continental shelves in many areas of our planet, represent a viable large-scale solution. However, we have as yet no technology for the recovery of methane, and methane is also six times more powerful than carbon dioxide, in terms of the contribution to global warming per kilo. The huge risk of methane emissions and the lack of an existing technology for the retrieval of methane largely inhibit the short-term pursuit of this alternative.

Instead, many experts agree that the solution most likely will consist of a number of different component solutions, which together will replace or upgrade our existing energy alternatives. However, even if many authors have analyzed the pros and cons of different alternatives and the volume of new power plants and vehicles or the state of sustainable energy technologies, nobody seems to hold an overall picture of how we will arrive at the energy system of the future. Increasingly detailed pictures of the future system are produced, for instance by the project Combat Climate Change, which involves a number of large companies, and there is a debate about the choice of alternatives.

There seems to be no clear route forward, which most experts are willing to agree upon to be the most preferable from a number of viewpoints. All alternatives have their drawbacks, and there are currently few sustainable solutions that have been implemented on a large scale. One of the main problems seems to be financing of many of the transformation measures. Despite their large financial resources, large companies, often legitimately, question the right of governments to impose substantial investment demands on them, which may reduce the competitive edge of companies, since they increase production cost.

We will here look at the different sectors where transformation will be necessary and also highlight some of the interrelationships between them.

Box 17.2 Wind Power – A Cost-Effective Alternative with Growth Potential

Wind turbines are currently one of the most cost-effective sustainable alternatives for power generation. New wind turbine projects are planned all over the globe, and the time from order to delivery of new turbines is currently a number of years. There is no sign that demand for new generating capacity will taper off in the near future.

By the end of 2002, the installed wind power capacity in North America was almost 5000 megawatts (5 gigawatts), compared to the installed capacity in Europe of some 23,000 megawatts (23 gigawatts). Still, wind power represented only 1 percent of the total electricity production in Europe and much less in North America.

In a 2002 report published by the European Wind Energy Association, this organization sets the target that by 2020, 12 percent of global energy production should come from wind turbines. To achieve this, 1260 gigawatts of capacity will have to be installed.[3] A large wind turbine has a capacity of more than 2 megawatts. This will require 630,000 new 2-megawatt turbines to be installed within the next 12 years, a truly astonishing figure. In the year 2002, less than 7000 megawatts, or 7 gigawatts, of generating capacity from wind were added to the global capacity. In order to add 1260 gigawatts until 2020, we would need to add 100 gigawatts per year for 12 years from now. While the production capacity for wind turbines is increasing incrementally (the capacity to produce new wind turbines has increased to 20 gigawatts annually in 2007), there is so far no sign that the dramatic scale-up of production capacity that will be necessary to meet the projected needs of the future is taking place. Also, the ambition to increase the average capacity of wind turbines has been tempered by the need to reconstruct subsystems, such as gears, to reduce the currently high rate of breakdowns and consequent high costs of maintenance. It is now a widely held belief by experts that the average size of new wind turbines will remain below 3 gigawatts for the foreseeable future, because the currently available components and subsystems are not sufficiently robust for larger turbines.

The huge investment represented by expansion plans will have to be managed in parallel with investments in other new and renewable technologies and production capacity in other areas. When viewed in this way, the sheer size of the figures indicates that the market will not be able to, without problems, allocate enough resources to all possible aspects of the transformation. We will need to prioritize alternative routes forward and drive planned and managed development in order to achieve both speed and reasonable cost-effectiveness.

As already argued, some very successful examples of planned industrial and economic transformations and development projects on a large scale are provided by the United States. Few other planned projects could match the ones represented by the US resources' build-up and transformation during three-and-a-half years from 1942 until 1945. The US space program is another example of a successfully planned and managed effort, which, among other things, put a man on the Moon in less than 10 years. Through The Marshall Plan, the United States provided financial resources and political help in order to create cooperation among the market economies of Western Europe. This greatly helped the economic recovery of Europe, and also benefited the United States, since economic growth created a strong demand for American goods.

In energy transformation, there will be need for a large variety of different technologies, products and services, in the same way that many different types of equipment were needed both during the World War and in the US space program. A planned effort does not rule out diversity in terms of the resulting products and services. Instead, a focused and planned approach, transforming a few major areas at a time, would generate faster results and be more cost-effective, than the unfocused, costly and scattered approach, which would inevitably be the result of a market-based approach. An approach that is, to a certain degree, planned is preferable compared to solutions that are entirely based on the play of the market forces. The degree of planning of the transformation process remains to be determined, and this decision needs to be based on a comprehensive analysis of our energy consumption, sector by sector. Only through this type of analysis could we arrive at a strategy and a plan for a cost-effective transformation.

Regardless of which alternative we choose, we need to look upon transformation as a large-scale program. We could divide the transformation program into six different areas, and organize work in each of these areas as a "stream" in the transformation program. Within each stream, we can then identify a number of different projects. The expansion of wind power generation capacity of a country to contribute to the possible global goal of increasing capacity in this area by 1260 gigawatts by 2020 could be one such project located within the stream of heating and utilities. Each country will have to run its program and its projects, and decide about its own approach to the transformation. Naturally, each country would also have to decide about the program organization and the way to divide the overall program into streams or subprojects. What is offered below could be viewed as an example and a start of the analysis effort.

The areas that could function as the relevant overall units of planning and management, and that could be turned into streams in the program, as illustrated by figure 17.1, could be

1. Transportation
2. Heating and utilities
3. Industrial processes
4. The built environment
5. Agriculture
6. Behavior and work life change

The approach above represents a manageable alternative to other approaches, such as the "wedges" proposed by other experts and politicians. The main difference between the approach presented in this book and the wedges approach is the ambition to manage the overall pace and manage change on an overall level, weighing efforts in different areas against one another. In the wedges approach each wedge is seen as separate from the others, making it possible for companies to work with development efforts in one wedge irrespective of efforts in

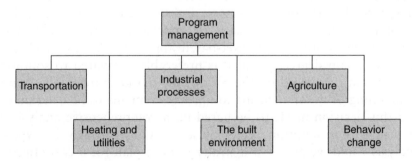

Figure 17.1 The structure of the program and the streams

the others. The ambition to manage the overall change is similar to that of the US government to manage resources' build-up for the Second World War or to manage the overall configuration of systems for the Apollo program. The wedges approach would not have been appropriate in either of these cases. It could suffice, however, if the financial and pace demands of the energy transformation program would not require any prioritization between areas (wedges or streams) or overall management of results against tighter time frames than the ones that could be achieved by the market forces left on their own. The wedges approach could be complementary to the approach proposed in this book or it could, under certain advantageous circumstances, be an alternative. Only a thorough analysis of the situation and the transformation needs can provide an answer to the question of which alternative is the most appropriate.

In the following chapters, the need for transformation and some of the opportunities in each stream will be discussed. If we run change in a clever way, we will be able to transform first those areas that are easiest to transform, and those in which we can achieve change at the lowest possible cost. Even if we can never know exactly which alternatives are the best for the future, we need to use the knowledge that we have, prioritize and make a plan.

CHAPTER 18

Transforming Transportation

The challenge of creating sustainable energy systems is greater in transportation than in other areas. Twenty-seven percent of US energy consumption is used for transportation and the sector is heavily dependent on a few very convenient and potent fuels that are based on petroleum as a raw material. Global oil production amounts to 85 million barrels per day. Ninety percent of global transportation is based on oil as the energy source and 60 percent of all oil that is consumed globally is used for transportation.[1] We have been able to pump oil up from the ground or the bottom of the sea in large volumes at low cost for more than 100 years. This convenience, and the prospect of continuing to increase petroleum production for more number of years into the future, has made us complacent. Many of us, however, are starting to see an urgent need to develop high-volume replacements. To many this need has previously not been as obvious as it is now, with signs of "peak oil".

In addition to the investments and changes that are necessary in order to develop new engines and power train systems, we will need to develop large-scale production facilities and distribution systems for renewable fuels. We may also need to transform existing large-scale production and distribution systems into a number of medium-scale systems, catering to smaller and more specialized segments of the fuel market.

Currently, global oil production and consumption amount to approximately 85 million barrels per day. Saudi Arabia is the largest producer in the world. The production by this country accounts for 11 percent of global production. Saudi Arabia and a small number of other countries of the Middle East represent some of the world's largest exporters of oil, and these countries are the only remaining swing producers that may still have some capacity to increase production in the case of a reduction in the production of other countries. It is, however, unclear how large this available capacity is for the short and medium term. This heavy dependence on a very politically unstable region of the world represents another strong reason for the large-scale development of oil substitutes. The uncertainty about actual remaining reserves, and the belief among a growing number of experts that the global peak in oil production may be imminent, represents another strong motivation factor.

Considering the very large volumes of oil that are consumed on a daily basis, this is an enormous undertaking. As Richard Heinberg has estimated, in order to produce sufficient amounts of biofuels from grain, to supply the US automobile and truck fleets, all the land in the country would have to be used for fuel crops, and there would be no land left to live on or to use for the production of food. This indicates that we need multiple routes forward, in order to transform energy systems for automotive fuels to sustainable alternatives. Most probably, we need to reduce fuel consumption, turn to electricity or fuel cells, since we can produce electricity from wind or photovoltaic cells, and combine these measures with large-scale development of cellulose-based fuels. We need to use our arable land for fuel production. Fertilizers and pesticides are produced using oil and natural gas as a raw material, which may represent both a problem and an opportunity. If oil becomes scarce we may have less of these. Furthermore, tractors and combine harvesters run on oil. If we have less oil, we may have to reduce the amount of work on the land by farmers, who use equipment for plowing or preparing the soil in other ways. We may reduce oil consumption by cutting down on the use of fertilizers, but then we need to run a large-scale change program in agriculture that may include large-scale research and information activities on large-scale farming using less fertilizers and pesticides. Some large companies, perhaps most notably IKEA, as described below, know how to run this type of program.

18.1   THE ORGANIZATION OF A TRANSPORTATION CHANGE PROGRAM

Due to the strong systemic links between vehicles, fuel and fuel distribution, a change program needs to coordinate efforts in the three fields depicted in Figure 18.1.

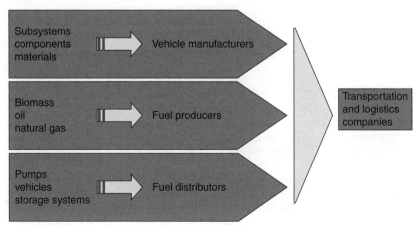

Figure 18.1   A change program for transformation of transportation needs to be divided into three large projects: vehicles, fuels and fuel distribution

In order for a change program to be successful, the organization would also need to include competence in the fields of change management and entrepreneurship. This is because all the products, components, materials and services that are necessary to go through with the transformation do not exist. Instead, new systems and parts of systems will need to be developed, sometimes from scratch, sometimes based on existing solutions and technology.

Due to the size of this sector, it will not be possible to transform the whole sector in one country, or internationally, in one single project or program. A project, at least initially, will need to focus on a subsector within transportation. This could mean that one project focuses on transforming part of the regional transportation in one region to the use of one particular renewable fuel. Other subsectors that may be the subjects of early projects may be city logistics in one or a number of cities, logistics for one industry, such as liquid food, in one country or other national transportation subsystems. It may be the case that international transportation will have to wait until a later stage, but regional or industry-oriented projects may still require international agreements about the way forward for, say, European transportation and fuels, in order to make major investments possible.

Below, as in figure 18.2, is shown an idea of an organization for a transformation program for a part of the transportation sector. In such a project new business development and entrepreneurship need to take a prominent role, because of the need to develop new components, systems and technologies and rapidly commercialize them on a large scale.

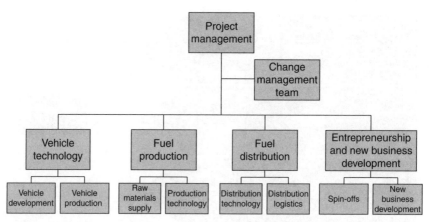

Figure 18.2 Suggested project organization for large-scale managed change project in the transportation sector

18.2 POSSIBLE ALTERNATIVE FUEL AND TECHNOLOGY PROJECTS WITHIN THE TRANSPORTATION STREAM

As we create the strategy for the transition, we need to look at both the long-term and the short-term perspectives. It will be easier, in order to achieve short-term change, to start change efforts by the rapid volume increase in the penetration of existing energy-efficient technologies. The easiest and least costly, short-term alternative is represented by the rapid growth in the use of hybrid engines, which is a combination of a diesel-, or gasoline-, fueled engine, and an electric motor driven by a battery, which could either be loaded as the vehicle brakes, or from the power grid during stops.

In the case of hybrid power trains, they represent a substantial opportunity to reduce petroleum consumption at short notice. At the moment, however, this technology is new and more expensive than traditional technologies, which reduces the speed of market penetration. As another alternative increasing amounts of some biological fuels could be mixed into the petroleum, up to a content of 30 percent, without any modifications on the vehicles. This represents an opportunity to increase the use of renewable fuels for older vehicles too. In addition to the need to build production capacity for hybrid vehicles, we would need to build production capacity for biofuels, and, possibly, increase the electricity generation capacity, in order to charge hybrid vehicles, and electric vehicles, from the power grid.

18.2.1 Hybrid Power Trains

Assume, for instance, that the goal is set that 25 percent of the energy consumption by trucks, buses and automobiles, by 2015, will be accounted for by vehicles with hybrid drivelines. In order to achieve this, substantial volumes of hybrid vehicles will need to be in production within only a few years. Since hybrid drivelines could be run using existing fuels, we would be able to reduce the consumption of petroleum-based fuels, while we wait for large-scale production facilities for biofuels. As mentioned above, using hybrid engines, we could also increase the use of biofuels, such as ethanol, methanol and biodiesel, simply by mixing them into petroleum-based fuels. In this way, we would not need to rapidly build new large-scale distribution systems for new biofuels. This would reduce both the risk and the necessary investment within the transformation program.

The investments, in this case, would require a large-scale effort to develop and perfect hybrid driveline technology and retool plants in order to substantially and rapidly expand production of vehicles using

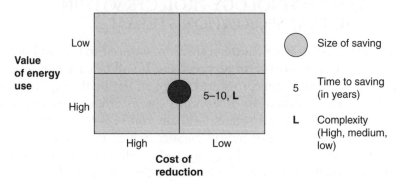

Figure 18.3 Diagram illustrating the potential savings, time frame, cost and complexity of a large-scale transformation project based on hybrid power train technology

this technology. Compared to the large-scale transformation of US industry during the Second World War, this seems to be a feasible task. Above, in figure 18.3, the project prioritization model introduced earlier in the book is used in order to roughly describe the time frame and the potential of the hybrid power train project.

The above matrix could be used to prioritize the hybrid power train project compared to other alternatives. Since hybrid power trains are already available on a small scale, the capacity to produce trucks using this technology can be expanded relatively rapidly should demand increase at the same pace. However, trucks with hybrid technology will be more expensive than existing alternatives and the penetration may be slow, unless incentives are put in place. In case only 5 percent of all trucks sold have hybrid power trains and 5 percent of all trucks are replaced in a particular year, this means that only 0.2 percent of the truck fleet are changed to hybrid trucks during a particular year. This means a very small impact on energy consumption.

The matrix also indicates that the economic value of energy used for transportation is high. This means that the cost to society of rapidly decreasing the level of transportation, without putting in place alternatives for communication or local production of food and goods, would be high. The size of the saving could potentially be high, as indicated by the size of the cycle, but the cost of reduction may also be high, if rapid penetration of the new technology is attempted. Hybrid power train technology is one of the immediately available technologies that could reduce the energy consumption in goods transportation and that could be implemented in a few years. Maybe we need to make the financial resources available that would make it possible to rapidly increase demand, or we may be forced to reduce transportation anyway, due to decreasing supply of oil in the not too distant future.

18.2.2   Electric Cars and Trucks

Assume also that a steering committee sets the goal of achieving 10 percent electric vehicle sales out of the total volumes sold in the automotive sector by 2015. In such a case, this could partly be achieved through increased sales of hybrid vehicles, which could also be charged via the electricity grid at night or during stops. The development of more powerful batteries for hybrid vehicles is underway. Demand, however, is still low. Increased demand would rapidly contribute to the development of increasingly powerful batteries for electric cars and trucks. Depending on the usage patterns of cars and trucks, vehicles that are destined for delivery and short haul freight, shopping trips and other short trips, in the case of cars, could increasingly be fitted with only an electric engine.

A rapid growth of electric vehicles would require a more complex set of actions. First, development in the automotive sector will also need to focus on the development of hybrids and electric engines with the ability to charge from the electricity grid. Second, in order to focus on the adoption of the new technologies by companies with large vehicle fleets, to create a sufficiently large market in a brief space of time, a sufficient number of charging units would need to be installed. Initially such facilities could be rolled out in the facilities of a few large, strategic users. These strategic users could be transportation companies, distribution centers, car rental firms or large companies with large fleets of corporate vehicles (trucks or cars).

Without a large amount of analysis, it also seems to be a good idea to promote or partly finance the replacement by electric vehicles of existing vehicles in companies with large car fleets and a high turnover of vehicles. These initial large-scale users could be car rental companies or large corporations, where a large number of vehicles visit central parking lots on a daily basis, where charging facilities could be installed. It probably makes sense not to focus on single unit users in the first place, because of the cost of installation and the larger administrative resources that are needed in order to market new technologies, support users, and administrate possible financial incentives, in complex networks of small users. A focus on small users would also, probably, be more risky, because it would be difficult to get the commitment of a large number of single-unit users to make the investments in vehicles with a new technology, and a number of other cost and risk items, besides marketing, would increase, compared to a focus on large customers.

Box 18.1 Advantages of Electricity

A former manager of R&D at ABB, Professor Harry Frank, argues that, in the face of declining resources of petroleum, we should use electricity for more purposes than the present ones. Electricity is a highly efficient carrier of energy and it could be used, among other things, to power cars and trucks. With increasingly powerful batteries it is now possible to drive relatively long distances, also at high speed. The capacity of batteries and related technologies is now increasing very rapidly and the time for charging is decreasing. ABB has developed an electric sports car with a range of 300 km and a maximum speed of upward of 100 km/h. These developments make electric cars and trucks more promising than fuel cells, because of the energy loss of some 60 percent in the production of hydrogen. Electric cars will be less expensive than fuel cell cars and they will require less expansion of energy production.

In addition to the investment in the development of technology for the charging of electric vehicles and the investment in the charging stations, during negotiations with companies that are targeted for early investments in charging units, steering committees need to take into account the consequences of the need for the large-scale charging of vehicles on electricity production and distribution. This, in addition to other changes in the demand for electricity and changes in the usage patterns of electricity, would need to be taken care of within the stream of energy production and distribution, outlined below. All in all, the goal related to electricity use for vehicles may be slightly more

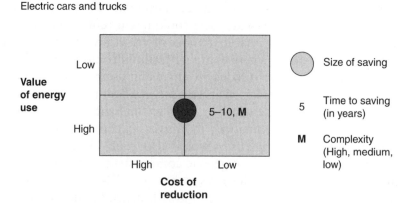

Figure 18.4 Diagram illustrating the potential savings, time frame, cost and complexity of a large-scale transformation project based on electric cars and trucks

challenging. In figure 18.4 above, the electricity challenge is roughly prioritized along the dimensions of the prioritization matrix. The prioritization in the matrix looks the same as in the case above, only that the complexity is medium instead of low.

18.3   BIOFUELS

The adoption of biofuels grows very rapidly, but from a very low level. Less than 1 percent of our energy needs are supplied by biofuels. There is substantial optimism regarding the future of these fuels, but a large amount of analysis remains to be performed in order to determine their role in the energy mixes of the near-term, medium-term and long-term future. As we will see below, there are biofuels with better prospects than ethanol that is currently expanding very rapidly in many markets globally.

Ultimately, the factor that will limit the adoption of a particular biofuel will not be the technological advantages of using it, but the availability of raw material, and the cost of production of the fuel, when it is used on a large scale. Another important factor will be the energy need in production, which is one determinant of the net energy that could be generated in each case.

In order to determine how much of our energy needs could be supplied by biofuels in the future, we need to make an overall map of the available alternatives, and their implementation, and decide which biofuels will have to be used in the future. Then we need to make conscious efforts to develop economically viable production technologies, engine solutions and distribution systems for those. Since it may take a decade to develop some of the production technologies, and expand production capacity for some of the alternatives, it makes sense to start, or prepare to start, before there is an urgent need for a new alternative. This probably means that we have to start "yesterday."

Although we have some renewable alternatives already at hand in small quantities, we will probably have to start parallel development or expansion of a number of alternatives. Exactly which biofuels would need to be developed will differ from one country to another, depending on short-term and long-term access to raw materials, the access to existing fuels and also depending on the price considerations.

What we do know for certain is that biofuels will never suffice in order to replace fossil fuels on a large scale. There are simply not enough of the raw materials to make this possible. As mentioned above, Richard Heinberg argues that all the land of the United States would have to be used for grain production if we were to replace petroleum by grain-based ethanol, and then there would be no land left to live on and no grain left to eat. Cellulose-based biofuels would give a better return, but there is not enough forest or land available for forestry to entirely replace petroleum consumption. In addition to this, gigantic logistics systems would be needed to take logs from the north of Europe or from Canada, or fuel produced close to the sources of raw materials to the markets where they are needed. Some analysts argue that we need to develop radically new technology, instead of relying on biofuels.[2] The exact nature of this technology is often not discussed, but we have to assume that some of the alternatives that may be considered may be among the ones mentioned in this book.

18.3.1   Grain-Based Biofuels

The primary use of grain in the future will have to be for food. We will not be able to set aside large areas of arable land for the long-term agriculture of energy crops. Due to decreasing global production volumes of grain, and the deteriorating growing conditions in many areas of the world, we may, instead, have to use more land for agriculture, without using very much of it for fuel. Grain-based ethanol could be used as an interim solution to create a demand for new alternative fuels, and as a "market maker" for these alternatives.

Through the increasing use of ethanol as a fuel for cars, either as an additive to gasoline, or as a separately distributed fuel, mixed with small amounts of petroleum, we have the opportunity to create a market for ethanol-fueled cars, and a debate around the use of ethanol.

There are other reasons to temper the hopes for a grain ethanol-based fuel economy of the future, besides the limited land resources. Grain is also produced in agriculture, which is a relatively energy-intensive type of production, compared to forestry. In many cases the net energy content of ethanol from sugarcane, corn or wheat is only some 10 percent of the total energy content of the fuel. This is because approximately 90 percent of the energy content has to be supplied in the form of petroleum-based fuels or biofuels in order to fuel tractors, combine harvesters and produce fertilizers and pesticides and other inputs into agriculture.[3]

According to some studies, it may be possible to reduce the amount of land needed for agriculture by some 15–20 percent, with improved methods. If this were the case, land could be used for fuel crops that would yield as much as 25 percent of the total global use of primary energy, which is about as much as is currently supplied by coal.[4] In order to achieve this, a massive project to reform agriculture would have to be run. On the other hand, ecologists argue that we need to turn increasing amounts of land to ecological agriculture, which normally brings smaller yields than industrial agriculture. This would require more land for food production instead of less and this would also require a large-scale project to reform agriculture. One intermediary solution would be to continue with industrial agriculture, but reduce the use of fertilizers and pesticides in order to reduce the use of petroleum for these purposes. This would also require a large-scale project to reform agriculture. Biofuels is certainly an area where the streams of transportation, energy production and agriculture intersect and the strategies chosen in each of the streams will influence the others. We also find that the demand for grain for biofuel production influences the price of grain in world markets. This is true already, when grain-based biofuels only supply less than 1 percent of our global fuels.

One more aspect of biofuels from agriculture, which is important when we consider making more land available for agriculture, is the aspect of binding carbon dioxide. In the future, we will still need to bind as much of the CO_2 emissions as possible through the growth of trees, shrubbery or grain production and we need large forests with large trees in order to consume as much CO_2 as possible from the atmosphere. For this reason there may be a limit to how much forest we can use in order to produce ethanol or methanol and how much land we could convert from forestry to agriculture.

From this perspective, a forest binds more carbon dioxide, than the same area used for grain production. At this point, some readers may argue that we also have the opportunity of sequestering CO_2 in underground deposits. Yes, this will be possible, but in a situation in which we have to use the available capital in the most productive way possible, sequestration should probably be seen as an alternative of last resort. The best alternative will be to increase the speed of development of energy generation technologies that are not dependent on CO_2 emissions. Sequestering does not produce any new energy, and it does not reduce our dependence on petroleum, which we need to reduce for other reasons than for the protection of the environment.

Box 18.2 Sequestration Technologies

Sequestration of CO_2 into large subterranean chambers is in itself a very large project. This alternative may be necessary to use in developing economies that need to expand coal-based electricity production. This is the case in Poland and China and in a number of other countries.

The cost of and time schedule for sequestration, from the perspective of the overall transformation program, should be weighed into decisions to expand the use of fossil fuels. This would have to be done in the same way that the cost of the expansion of nuclear power generation needs to include the cost of the final storage of the used uranium or plutonium. The cost of sequestration will then have to include the cost of solutions for the transportation of CO_2 to the deposits that will seldom be located next to the power plants.

The sequestration system will have to include facilities for the storage of CO_2 at the power plant, vehicle fleets (or pipelines) for the transportation of CO_2 from power plants to the sequestration facilities and for the sequestration facilities themselves.

The alternative will be to use the capital available to invest directly in the development and expansion of production of sustainable energy systems, using renewable fuels. At this point it is too early to discount any possible solutions, but we need to take into account that it will be several years before we have access to full-fledged systems for storage, transportation and sequestration of carbon dioxide, and at that point in time, we will have less oil to bother about, and we will already, because of the peak in oil production, experience a reduction in carbon dioxide emissions from cars and trucks.

This is not to say that sequestration will not be necessary – only that we will need to see the need for sequestration in the light of the necessary reductions in petroleum consumption in the near- and medium-term future and prioritize between investment needs based on a complete picture of the future situation. In order for this picture to be complete, we also need to take into account the fact that the peak in coal production is also likely to occur within the next few decades. While there are still large amounts of coal in the ground, the deposits that are easiest and least costly to develop have been taken advantage of. In the future we need to turn to deposits that are more costly to develop and this will gradually reduce the amount of coal produced. Once more, even though there are plans for the large-scale expansion of coal energy in the near future we need to base sequestration plans on the actual volumes of fossil fuels that will be used when the sequestration facilities are ready. CO_2 emissions will probably start to decline sometime within the next decades and we will be forced to replace these technologies with sustainable ones.

18.3.2   Cellulose-Based Biofuels

As indicated above, we need to base an increasing amount of our future energy needs on biofuels, produced from cellulose. We already possess a number of alternative fuels that could be used, using different technologies, from distillation of ethanol, to the production of a number of different hydrocarbons from wood. In the case of methanol, we already have a process, which is available for the large-scale production of this product. As mentioned already, methanol can be used as an additive to petroleum. Similar to ethanol, methanol can be added in quantities of up to 30 percent to gasoline, without requiring any modifications of the engines.

In the case of ethanol, we already have some global production of this fluid, based on grain as the raw material, but at present, we do not have an economically viable process for the production of ethanol from cellulose. This process has been under development for decades, without success, and we do not know when one will be available. In the energy transformation it will be important for a steering committee to understand why this effort, so far, has failed, and what the prospects for success are in the near future. If we want a cellulose-based additive to gasoline, methanol may turn out to represent a more rapid path toward sustainability and financial gain than ethanol.

All in all, there are substantial resources in our global forests, both in the form of wood and waste from forestry. One of the problems is that the raw materials are scattered across large areas and that there is a high cost, both in terms of energy and money, involved in collecting it. Based on rough estimates, the available waste from forestry amounts in terms of energy to one-sixth of the total current primary energy supply from oil, coal and biomass. However, the net energy content is only 30 percent when the energy needed for the production has been subtracted. In addition to this, this energy is spread out across the forests of the world. In order to produce this energy in an efficient manner, we need to build a number of plants for biofuel production that are close to the forests and we need to invest in resources for the collection of the waste materials on location.

Collecting raw materials is not the only challenge. In order to produce fuels from cellulose at a scale that would be financially justifiable, each plant would need deliveries of 450 truckloads of wood per day. The number would be higher if we plan to use waste materials from forestry for production. These loads would have to be supplied by truck, because forests in general are not located along train lines or close to harbors. They are also spread out so that large railway systems would be needed in order to serve a large forest area, which may not be a viable solution compared to truck transportation. In order to produce enough fuel to supply Poland with 20 percent of its current fuel needs, three plants of the mentioned size would be necessary. In order to supply the 15 largest economies in Europe with 15 percent of their needs, 122 plants would be needed and in order to supply 100 percent of European needs 1000 plants would be needed. Since most of the forests in Europe are located in the north, this type of system would present a tremendous challenge from a logistics perspective.[5]

Box 18.3 A Comparison between Different Biofuels

The global leader in the truck industry, Volvo, sees the transition to renewable energy sources as a critical move to ensure the future sustainability of transportation. In early 2007, the company announced plans to launch a range of prototypes for hybrid and biofuel-powered trucks. This is only one measure taken by the company in order to contribute to the development and large-scale implementation of sustainable technologies for transportation.

According to Volvo CEO, Leif Johansson, what is now needed is a large-scale development effort to produce and distribute renewable fuels. He also calls for coordination between producers and legislators, across national borders, in order to ensure a long-term commitment to a universal set of fuel standards for the future. In order for Volvo to succeed in becoming one of the solutions to the climate problem, and participate in the development of carbon dioxide–free transportation technologies, broad agreements at a high political and industrial level are necessary.[6]

In a booklet published by Volvo in the spring of 2007, the company published an analysis of the road forward in the case of carbon dioxide–free and renewable fuels. In the booklet, the advantages and disadvantages of each of the following fuels are recounted in a brief and very accessible way.

- Biodiesel, which is diesel based on vegetable oil, such as rape and sunflower seed oil.
- Synthetic diesel, which is made through the synthetic production of hydrocarbons, based on the gasification of biological material.
- Dimethyl ether (DME), which is a gas, which could be distributed as a fluid under low pressure. DME is also produced from gasified biological material.
- Methanol, produced through the gasification of biomass, and ethanol, produced through the fermentation of vegetables that are high in sugar and starch.
- Biogas, which could be produced in several ways from biomass from sewage treatment facilities, and at waste deposit stations,

and in other places where biological material can be gasified. Biogas needs to be compressed to 200 bar, and then it requires the engine to be fitted with a spark plug, which gives it lower energy efficiency than the other alternatives.

- Biogas and biodiesel in combination, which are stored in separate tanks, and injected by different injection systems. A small amount of biodiesel is used in order to ignite by compression. In this solution the biogas is kept in a cold and fluid form.
- Hydrogen gas and biogas in combination. The hydrogen is added in small proportions (in the case studied by Volvo, 8 percent) to compressed biogas. Hydrogen could be produced through the gasification of biomass, or through electrolysis of water, using renewable electricity. This alternative requires the engine to be fitted with a spark plug.

In the study performed by Volvo the seven alternatives were compared across seven criteria related to their development and use in future energy systems:

1. Effects on the climate
2. Energy efficiency
3. Efficient use of land
4. Fuel potential
5. Necessary adaptation of vehicles
6. Fuel cost
7. Infrastructure for distribution

In the study, each fuel was evaluated, using a five-degree scale, for each of the criteria.

The renewable fuel that came out as the most successful alternative of the study was DME, which received high ratings along all criteria, except infrastructure. In the case of fuel cost, the rating is dependent on the production process. Gasification, in order to produce methanol and DME, is relatively expensive, compared to alternative production processes.

The infrastructure, Volvo remarks, may be important in the short run, but it does not represent a problem in the long run, since the infrastructure for gasoline and diesel also needs to be maintained and invested in. However, it is clear that the investment in new infrastructure will need substantial investments and that this needs to be taken into account in the program, since money will be a scarce resource.

Needless to say, all alternatives could be used in combination with hybrid power train technology. In this way, the energy generated as cars and trucks brake is conserved and reused in order to fuel the vehicles. Thus, all investments that we make in the further development of hybrid engines, the expansion of production resources and the sales of vehicles equipped with hybrid engines will pay off into the distant future in combination with a number of different types of fuel.

This study, and other similar studies for other areas, will provide important guidelines for the transformation program. However, we must not draw the conclusion that we need to identify the best alternative, and then apply it to 100 percent. In situations where we could use a particular raw material for a number of different fuels, we need to select the best alternative. We do, however, possess a number of different raw materials and we sometimes have access to those in medium-sized quantities as by-products from various production processes. Black liquor is one such by-product from the cellulose industry, which could be used as a raw material for fuel production. In such a case, fuel plants could be located close to cellulose factories, to take advantage of the synergies. In the case of hydrogen and other gases, which are necessary in the production of synthetic diesel, they could be found as by-products from different chemical production processes.

In the future, we will need to make use of a number of different energy sources and fuels, and we may need individual distribution infrastructures for each of them. Along the line of argument of this book, the transition from the current situation to a future situation, using renewable fuels, will require an overall plan and a system of goals and means to reach the goals, which can be communicated and worked toward by all the people who need to participate in the change program. As noted above, the CEO of Volvo, Leif Johansson, agrees.

Instead of drawing one prioritization matrix for each of the biological fuels, I draw one that may represent the biological fuels taken together (see Figure 18.5). This is done in order to illustrate the relative project status of biofuels compared to hybrid power trains and electricity. Below, matrices for all alternatives in the other streams will not be presented. A method, such as the prioritization matrix, needs to be used in order to compare different alternative projects with one another, but at this point we have too little information available in order to make this evaluation for all alternatives. This needs to be done as part of the strategy development phase of the transformation program.

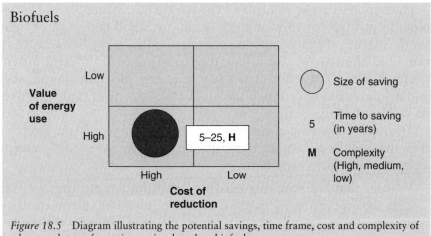

Figure 18.5 Diagram illustrating the potential savings, time frame, cost and complexity of a large-scale transformation project based on biofuels

Even though states may decide to not actively provide government financing of projects in a particular area, the minimum requirement of the program is that the steering committee of each stream define the long-term goals for different renewable energy alternatives. As argued by Volvo, this is the minimum requirement in order to make it possible for Volvo and other truck and car companies to become part of the solution to the climate issue.

18.4 COMPLEXITY REDUCTION IN TRANSPORTATION SYSTEMS

We will need to use a number of different fuels, production and distribution systems in parallel in the future. This may sound as a more complex situation, which will be more difficult, costly and risky to handle, than the existing system of gasoline and diesel. I do not argue in favor of a number of different solutions, for the sake of increasing complexity. We need to reduce our reliance on fossil energy systems, and we will not be able to satisfy the future needs of the transportation sector, using only one fuel. Instead, we will need a combination of hybrid solutions, electricity and renewable fuels, in order to secure the future supply of energy for transportation.

In the transformation effort, we need to take capital investment needs into account, as we make our plans, but we will have to involve a number of different companies and other players, in the effort, in order to make rapid progress possible. How can we manage this process, without making it so complex that it becomes impossible for people to make decisions?

Once more, I argue in favor of the advantage of analysis and planning. In this type of very complex situation, it will take a lot of time for the market, through experimentation, to identify the best combinations and alternatives. For a long time, companies in different industries will have to experiment at a low level with different alternatives, and it will remain uncertain if the small-scale experiments are going to be viable on a large scale.

Through a planning effort, we will be able to identify the alternatives that will be best suited in particular situations. We could identify a number of typical situations:

- Vehicles that are primarily used for local short range trips by car. Cars that are primarily used for such trips could be run on electricity, or on some locally available fuel. When used for local driving, hybrid cars, or all cars with chargeable batteries, could be charged from the electricity grid. When used for driving longer distances, hybrid cars could use renewable fuels in combination with electricity. It needs to be emphasized once more, however, that the large-scale use of electricity for transportation will require massive increases in electricity production.

- Primarily local short range trips by delivery trucks. There are a number of different local and medium-range uses for trucks, ranging from deliveries of parcels, to delivering gravel, ready-mix concrete, beverages and other locally and regionally transported goods. There are also local services that require vans or small trucks for transportation of service technicians. Trucks and vans for these purposes could be fitted with electric motors, or hybrid engines, in combination with renewable fuels. As increasingly powerful batteries are developed, heavier trucks could be fitted with batteries and transportation could thus gradually be transformed to electricity.

- Companies with large vehicle fleets could specialize in the use of a few particular renewable fuels, in combination with hybrids and electricity. Vehicles with different fuel systems could be used for different purposes.

  • In the case of local transportation or for the use by service companies, the facilities for refueling could be located at one place. In the case of companies with smaller fleets, a network of refueling stations in a region, or locally, could be developed for a number of companies by a fuel distribution company.

  • In the case of long-range transportation, fuel stations could be distributed along major routes, or located at the distribution centers run by the transportation companies. This is not very different from the present situation, in which many transportation companies have their own refueling facilities for their fleets.

Most probably, it makes the most sense to focus on marketing the renewable fuels, and vehicles that use them, first to companies with large vehicle fleets. This will be in order to reduce the capital needs of the project, both for investments in the distribution systems and in the cost of sales and marketing of these systems solutions, which will increase for smaller customers.

18.5   SHORT-TERM AND LONG-TERM SOLUTIONS

18.5.1   Fuel and Engine Technologies – Short-Term and Long-Term Solutions

We may view solutions that can be developed and somewhat expanded from now and over the next 10 to 20 years as short-term solutions. We may look upon solutions that may take more than two decades to expand into high-volume alternatives as long-term solutions. Within these broad categories there are, as has been indicated, subcategories of technologies that are already available and others that need some years' further development. In some cases, we may be able to continue to expand short-term solutions, so that they can cover an even larger share of our energy consumption in 50 years. It may also be the case that some solutions have limitations in terms of how far we can expand them. This may be due to limited land for agriculture and forestry or limits to other resources. In order to avoid wasting money through lack of planning, program priorities need to focus on solutions that are viable long-term. In order to be able to start to reduce energy consumption rapidly, using solutions that may not be possible to expand very far into the future, we may still have to employ solutions short-term, which we will have to gradually abandon within a few decades. In the case of long-term solutions for transportation, it is possible that the supply of grain and wood may limit the opportunity to expand production of biofuels beyond a certain point. It may still be necessary to develop these solutions up to a point.

It is important to note, as has been done above, that hybrid engine technologies and battery-powered vehicles represent solutions that already exist and that could be used for a long time into the future, provided that we manage to expand electricity production. Other technologies we have less knowledge about. In the case of grain-based ethanol, the expanding global population may need to eat all the grain that we can produce in the future.

18.5.2   Fuel and Engine Technologies – Visonary Alternatives

The solutions that we have already discussed as short-term and long-term solutions represent alternatives that have been researched and developed to a substantial degree, so that we have reason to believe that these technologies will represent viable solutions for the short or long term. In addition to these technologies, there are a large number of creative ideas of which we do not have much knowledge. We do not know if they are going to work on a large scale. They may be ingenious, but we also have to assume that several of them, for different reasons, will not work, even though they seem promising at the prototype stage. In other cases, it may be that the solutions are simply not attractive enough to present-day customers, but that changing circumstances of the future may change the preferences of consumers and lower the thresholds for the development of creative new technologies.

In the automotive industry, such technologies are sometimes exhibited at car shows, as concept vehicles. In this book such technologies are called "visionary" technologies. The suggestion is that we, as a society, should apply similar methods for prioritization between technologies that well-run companies do when they analyze the opportunities, potential market size and risk of different alternatives at an early stage, based on a small volume of experimental data, market data, risk analyses, and so on. Then they prioritize the alternatives and select a very small number of the more promising ones, and make plans for the development and commercialization of these alternatives, based on realistic plans, in order for them to make it all the way to the market and mass production. In the same way the managers of the transformation program should evaluate visionary technologies and select a small number for further development, and spend sufficiently on those in order to help them into the commercialization stage. The need for investments in technology development depends on which other investors are willing and able to invest. In some cases a visionary technology may be supported by large companies, which are willing to invest substantial sums in the development and commercialization of the technologies. In other cases the technologies may be developed and owned by private citizens or small companies who have limited capital to invest. These two basic types of alternatives need to be treated in different ways by the program.

Box 18.4 The Air Car

In the future, cars may be fueled by air. The French engineer Guy Negre used to be a designer of Formula 1 racing cars. During the past 15 years, he has been working on the construction of a passenger car that is powered by compressed air.

Although this car could be described as a visionary vehicle, marketing of the car were to be started in 2008, when 6000 units were to be made available in the Indian car market!

Reportedly, the lightweight chassis makes it possible to drive 125 miles on a single tank of air, and it only takes a few minutes to refill. In one version, the car is also fitted with a gasoline engine, which powers an onboard air compressor. This increases the range of the car, so that the full distance from Los Angeles to New York City can be covered on one tank of gasoline.[7] If this technology proves to be as good as it sounds, it may become a serious competitor of the combustion engine.

When the plans for some such technologies have been made, it can be determined whether they represent opportunities for the short or long term. Needless to say, the time to market this type of technology ought to be an important factor to weigh in the decision about government funding or support. In a market economy it will always be possible for an inventor, entrepreneur or a company to invest as much money and resources as they wish in a project that they believe in. At the level of society, and regarding projects that qualify for a planned program, the selection, in order to reduce the number of projects and the volume of investment, needs to be as strict as it would have been in any company with a management that manages toward the return on investment of its shareholders. This means that the planned project needs to be run toward the goal of supplying the energy needs of the future through the development of as few new technologies as possible.

Box 18.5 Algae Powering Fuel Cells

Hydrogen is seen as the ideal fuel of the future by many, because it is the most abundant chemical element on earth. Hydrogen can be used to power cars through fuel cells. The production of hydrogen from water through electrolysis requires, however, large amounts of electricity. Some experts claim that the production of hydrogen from water requires three times more electricity as the amount needed to drive an electric car the same distance. On the surface this would be an argument in favor of battery cars compared to fuel cells.

Researchers of the University of California at Berkeley now investigate the opportunity to use green algae in a pond of water to produce hydrogen. When sulfur and oxygen are withheld from the algae, they produce hydrogen gas. According to the scientists, one small pond could supply the fuel needs of approximately a dozen automobiles on an ongoing basis.[8]

If we turn to the case of the Second World War technology, the United States did develop a limited number of models of aircraft, tanks and other equipment. The actual number was based on the need, rather than the idea that different alternatives are needed in order to compete with each other. In the case of the space program, any rocket or space shuttle will only need to have one technology, or in some cases a small number of technologies that complement or work in parallel to one another, for a particular purpose. The number of technologies that have to be developed is based upon the need. In all cases, decisions about which alternatives to develop are taken at an early stage, so as to spend as little money as possible on alternatives that will not be needed. In the case of visionary technologies, it is also important to sort out less promising alternatives at an early stage.

18.6 TRANSPORTATION SYSTEMS

Using new fuels and engine technologies within existing transportation systems will reduce energy consumption and emissions of greenhouse gases. It is, however, possible to create new transportation systems and increase the efficiency of existing ones. There are many possible systems changes that could be made. Following the argument earlier in the book, we will have to prioritize the ones that could provide a substantial saving in terms of energy, require a limited investment per unit of CO_2 saving and leave savings that require larger investments per unit to a later stage, when we have already gone through with a number of highly investment-efficient early improvements.

We have already argued that the relatively low cost of fuel has contributed substantially to our present high energy consumption. This is true both for the transportation of goods and people. In the case of goods, the structure of global production systems has been established mainly based on the need to save labor cost. For most high- and medium-value goods that are low in weight and volume, even the remotest production location is not too far away to justify transportation of raw materials and components to and from countries with the lowest labor cost. In many cases, as we all know, the contents of products are often transported several times the distance of the earth's circumference.

As individuals, the low price of gasoline has increased the use of cars and it has also influenced the way that we build cities, plan housing and shopping and the way we commute to work. In all Western countries, we have substantially increased our amount of driving and we also use the car in situations where we previously went on foot or bicycle.

Increasing fuel costs are now threatening to reduce economic growth in global markets and no country is impervious to this development. We can also notice that our own consumption habits have had a substantial impact on the climate. In any effort to reduce energy consumption, decision makers need to take into account opportunities to reduce energy consumption in these areas as well.

18.6.1   Switch from Trucks and Airplane to Trains and Ships

It is to some extent possible to switch transportation from airplanes and trucks to trains and ships that often use less energy. In countries where trains run on electricity from renewable sources, such as hydroelectric plants, biofuel-based plants for cogeneration and wind power turbines, trains are much cleaner than all other alternatives. However, both trains and ships already operate at, or close to, capacity, and we need more trains and ships in order to switch large volumes of goods from trucks and airplanes. At some point in the expansion of train freight and shipping, we also need to increase the capacity of ports and railway systems. This is already the case for many railway lines that run at or close to capacity. In Europe one important measure to take is the harmonization of the different national railway systems, which could increase the efficiency and capacity of train freight substantially.

However, there is also the opportunity to make ships, trains and airplanes more fuel-efficient and less polluting. In the case of trains and airplanes, new generations of diesel engines for trains and new engines for airplanes, developed by GE in the project Ecomagination, cut some 15–20 percent of the fuel bill for these alternatives. However, according to the Hirsch report, the median lifetime of air carriers is 22 years, which means that current improvements in technology only slowly will make an impact on the energy efficiency of air transportation.

> ### Box 18.6 Magnetic Levitating Vehicles (Maglevs)
>
> One creative idea for long-distance transportation that needs to be further investigated is the Maglev solution, consisting of high-speed levitating vehicles that are propelled along a track by magnetism. This solution has been proposed for transportation of people and goods between Europe and China and the United States and China (via the Bering Strait). These vehicles travel at a speed of 340 miles per hour and the system, because of the one-way magnetic tracks, would be very safe.
>
> This type of transportation, when in place, would dramatically reduce the energy need for long-distance transportation compared to airplanes and existing train connections. It would make long-distance transportation possible, even in a future with access to much less energy than we have at present. The estimated cost of building this network is some 400 billion dollars, which is a huge sum of money, but still much less than the estimated cost of the Iraq war.[9] The researchers John B. Kidd and Marielle Stumm mention a number of different sources of financing that would be possible for this type of project. Among them is financing from nations with large amounts of foreign reserves, such as China, and from Sovereign Wealth Funds.

18.6.2 Short-Term Improvements

To some extent, we have the opportunity to use transportation resources more efficiently. In many cases, trucks run empty on return trips, because of the difficulty of selling free space on a return journey. Other contributing factors are the tight supply schedules of lean companies that require just-in-time delivery of goods. Up until now it has been seen as less important to save money on transportation than to reduce the lead times of production and logistics processes. This makes it more attractive to buy transportation space at full price than to use low-priced empty trips. One of the short-term savings opportunities for a steering committee in the transportation stream could be to set a goal for better utilization of transportation space. Using the line of argument that has already been used for behavior change earlier in the book, it is less clear than in the case of hybrid drivelines and renewable fuels, how such a project should be run, and which tools that are available to a steering committee or project management in order to effectively change the behavior of transportation buyers. Behavior change is clearly, to some extent, a matter of providing information to transportation buyers and transportation companies. This information could both be focused on the savings potential, in terms of CO_2, which could arise from a higher fill rate in transportation systems. It may also be a matter of creating incentives for companies that contribute to the achievement of better fill rates in transportation. There are Internet sites, where free transportation space is marketed. Maybe, stronger marketing of these sites would improve the situation.

It becomes increasingly obvious that changes in behavior in different parts of society will require a number of information, marketing and incentive campaigns, aimed at well-defined segments of transportation markets.

18.6.3   Change in Supply Chain Structures

Most probably, the cost of transportation will continue to increase, largely because of increasing fuel prices. This will gradually alter the priorities made by industrial companies that will increasingly look for less energy intensive transportation, or an increasing share of local and regional sourcing. Based on the description of the development of existing production and distribution structures the investment needs will be high and the process will most likely be slow. For the foreseeable future most companies will still find that there are substantial cost advantages in low-cost countries and the financial incentives to move production would be low. The available capacity for production in Europe and the United States will also be low in most cases.

The most rapid opportunities to reduce the amount of energy used in goods transportation would be related to reducing transportation of low-value goods, for which the transportation cost represents a substantial share of the retail price. This is true for agricultural products that are produced in hot climates and transported for consumption in temperate areas. Transportation costs sometimes amount to as much as 15–20 percent of the retail price of fruit and vegetables, which will increase to an even higher percentage, if fuel prices continue to increase. Still the price increase alone is unlikely to cause large-scale changes in consumption patterns. Again, information and marketing of desired changes may be more viable alternatives. The transportation of high-value goods represents a substantially smaller share of what the customer pays for the item and then the probability that a price increase will have a significant impact on sales will be smaller. In the case of a mobile phone, the transportation cost will represent 1 or 2 percent of the price. For inexpensive clothes, the share will be higher, but the cost of producing them close to US or European markets, using high-wage production personnel, would make the products even more expensive.

Changes in purchasing strategies and sourcing could contribute to the reduction of energy consumption. It is, however, not clear how much energy could be saved. One of the means available for such a project would be putting energy consumption for production and transportation beside the price on the price tag in stores. In such a case, energy-conscious consumers could use this information to increase their purchases of less energy-intensive products and reduce their purchases of products that need large amounts of energy for production and transportation, or increasingly purchase local or regional varieties or brands of these products, if they are available or as they become available.

A certain amount of energy could be saved simply by reducing the amount of purchases of imported goods and goods that have been transported long distances within the country. This would, in effect, mean that we were shifting products, so that people in each part of the country increasingly consumed the goods that were produced there. There will be a limit to the short-term opportunity to do this. Consider the brewery structure of the United States. There are now local breweries and pub breweries in many cities, but these have limited capacity, and most of their products are already consumed locally or regionally. If people want to continue to drink beer, they will have to continue to purchase products from large-scale breweries for years to come, which have to be transported some distance, from places like Milwaukee, Copenhagen or Amsterdam, depending on where you live and which beer brands you prefer. It will not be until a sufficient number of smaller breweries have been established, that this industry could, once more, become local. Yet, if all of us did buy products from producers located more closely to ourselves, it would still have a substantial impact on energy consumption. It would dramatically change distribution structures and remove incentives for companies to increase long distance transportation. The same is true for clothing. People may be able to buy some of their clothing needs from regional manufacturers, but it is not likely that the present structure of the clothing industry will offer too much opportunity for this. The available manufacturing capacity in most Western countries will be too small in order to supply the population with clothes produced in the nation.

We have moved most of our production capacity in many sectors to low-cost countries, and in the more developed economies people increasingly work in high-value service sectors rather than in production. In order to reduce transportation and emissions, we would need to establish more production capacity in our own countries, but we lack people with the necessary training and motivation to work in production, and the economic system forces our labor forces to move into trades of increasingly high value. As we see, a change to local or regional production will be difficult for a number of reasons. The price consciousness of consumers is only one aspect of the problem. There are a number of others that seem to pose equally difficult obstacles.

Unfortunately, it is probably the case that large-scale changes of supply chains will require very large investments in new production capacity and that we would need to make substantial changes to our economic systems in order to increase the supply of production labor in our present service economies. Because of this we need to count with very long lead times for large-scale changes to take place.

18.6.4   Transportation of People

For people, transportation is largely a matter of convenience. We all want rapid and easy commuting to and from work and we want the same when we need to go shopping or run other errands. Changing habits is often difficult and, in the Western world, we are not used to limiting the freedom of choice for people. There are also difficulties in the implementation of public transportation networks, based on the varying population densities of cities in different parts of the world. In order for large-scale public transit systems of buses or trains to be viable, we need a certain population density in the urban areas, where they are going to be implemented. Many US cities lack the population density, which is necessary in order to capture a large share of public transportation needs within a system.

In principle, there are two types of cases, one in which the population density is high enough to make a substantial increase in the volume of public transportation solutions possible at "short notice," the other being the case where a number of other changes, besides the implementation of a public transit system, will be necessary in order to facilitate the change. In this case the term "short notice" must allow for the planning and construction or the capacity increase of an existing system, which may take anything from a few years to close to a decade, depending on the situation.

18.6.5   Public Transportation in Densely Populated Areas

In many European cities, public transportation has been well developed for decades. The urban transit needs have been solved through a combination of trains, subway systems and buses. Through a building structure consisting largely of apartment and office buildings in city centers, population density is high and many people live within easy reach of a train station or a bus stop. In this type of city the public transportation system can be used for most purposes. Besides this, in many cities, a system of foot and bicycle paths and sidewalks are in place, which can facilitate short-range transportation that is entirely manual and thus CO_2-free. People who live a short distance from a bus stop or a train station may decide to walk or go by bicycle to reach the nearest node of the public transportation system.

In this type of situation, it is possible to rapidly add more bus or train lines, add more vehicles and increase the frequency of departures, and in other ways makes the system even more attractive and accessible to users. Shopping for food and other items is a major need that creates a need for transportation. In many countries, shopping malls and supermarkets have largely been built outside of the city centers, and the habit of doing weekly shopping rounds, instead of shopping on a daily basis, has been developed. This sometimes precludes small-scale shopping, because people need a car in order to bring large volumes of goods to their homes. In the process, the old neighborhood stores, which used to be located "en route" to and from work, have often been closed down, which also adds to the need of car transportation for shopping purposes.

In these situations, a number of different measures can be used in order to add to the attractiveness of public transportation:

- Increase the number of bus, train and subway lines of the systems.
- Increase the frequency of departures.
- Increase the population density of cities, by planning the construction of new office buildings and new houses within existing city limits.
- Increase access to the transportation system through better footpaths and bicycle paths.
- Increase the volume, accessibility and safety of car and bicycle parking at bus stops and train stations.
- Increase the number and size of local stores that are close to bus stops and train stations.
- Improve public transportation to and from shopping malls and business centers.

In conjunction with measures to make the public transportation systems more attractive, it is possible to restrict access to certain parts of a city by car, or by particular types of car. This would require improvement of access to city centers through public transportation. It is possible, however, to restrict access to city centers for vehicles run on fossil fuels, or to implement road tolls for cars and other vehicles that want to get access to city centers.

18.6.6   Public Transportation in Less Densely Populated Areas

Previous attempts in the United States to build large-scale public transportation networks have largely failed, due to lack of demand when the facilities were ready. This was the case with the Los Angeles subway. This project was terminated before the system was completed.

This is an example of a failure of a public transportation project, and it is important to keep such failures in mind, when new projects are planned. However, we must not rule out public transportation systems for US cities. We only need to keep the known limitations in mind, so that we avoid repeating past mistakes.

Changing transportation patterns is also a long-term commitment, and the implementation of a large-scale transit system should not be seen as a short-term effort. As has been touched upon above, transportation needs are intimately linked to the lifestyle as a whole. What are people used to do on their way to and from work? How could they organize their lives in a new way, using public transportation, instead of private cars? Which other aspects, besides the communication network, are needed in order to create a need for the new network, and offer a complete solution to the needs of a large share of the population? These and other questions need to be answered in order to understand how transportation patterns could be changed.

Box 18.7 The Case of the Öresund Bridge

In the year 2000, after more than a century of discussions between Denmark and Sweden, and a construction project spanning several years, the 20-km Öresund Bridge was opened. This bridge connects the capital of Denmark and the surrounding area with over two million people, with Malmö and a number of smaller cities in Sweden, with, altogether, one million people. The project was a tremendous engineering achievement, consisting of the construction of one entirely new island in the strait of Öresund, a 5-km tunnel and a 15-km motorway and train bridge.

This, as it happens, is not an example of improved sustainability, but of changes in communication patterns that take some time to achieve. The experiences of this case could be used in order to change travel patterns and improve sustainability in other parts of the world.

After the bridge was opened, the number of crossings was at a disappointingly low level, and the tolls charged to travelers made too little money to pay back the investment. Business between the countries, despite the regions' similarities, had never been particularly busy, but still many companies had activities on both sides of the strait, and many occasions for business travel had been expected. The international airport of Copenhagen is located at one end of the bridge. This airport is used for virtually all international business travel from the region, also by Swedes, so the traffic to and from the airport added a substantial number of transits over the bridge.

After a few initial years of disappointment, the number of crossings has now rapidly picked up, and in 2007 they amount to 18,700 vehicle crossings per day, up by 18 percent compared to 2006. At the same time, train traffic increased by 25 percent in 12 months, to 7.1 million passengers. 2007 is the first year for the bridge consortium to turn a profit. The project had budgeted for a few unprofitable years, before traffic picked up, and traffic did start out at an unexpectedly low level.

The most important reasons behind the rapid increase in crossings are probably the following:

- Increased integration between businesses on both sides of the strait. Companies that do business in both Denmark and Sweden used to have entirely separate organizations and distribution structures, which are now rapidly becoming more integrated, and this increases the need of transportation between the countries.
- People have started to take advantage of the differences between the two countries that are connected by the bridge. Many people in Sweden have taken jobs in the rapidly growing labor market of the Danish capital. Many Danes have moved to Sweden, where it is still substantially less expensive to buy a home, than in the crowded Copenhagen area. Many Danes had previously never been to Sweden, and they now discover that prices are lower and that the savings made on a purchasing trip can both pay for the crossing, and for additional shopping and meals.
- Many companies on both sides see the other side as a potential market for business expansion and start to visit customers and plan the establishment of business on the other side.

This is an example that shows the structural changes that need to accompany a large-scale transportation systems project. Before the bridge, it took at least 1 hour, but often 2 hours, for any trip in one direction, including boat and local transportation from the boat, and transportation solutions for visitors and people to and from work were not very well developed. Due to the long transit times between the cities on both sides of the strait, there were few links between people and businesses in the two countries.

Now, many people can board the train close to their home and get off a short walk from the office on the other side. It takes some time until large numbers of people change their lifestyles, and adjust a number of patterns of their lives to the new opportunities. The same must be true in areas with low population density, and where people are used to going by car all the time. It will have to take time for many people to change their habits and identify new shopping opportunities and places to live and work.

It must be emphasized in this context that it has taken between one and two decades from the decision to build the bridge to arrive at the present changes in lifestyle patterns. As an increasing number of facilities are built close to the bridge, and as a new train connection from the center of Malmö is completed, the integration process in this large city area may continue for several decades. Infrastructure projects are thus long-term investments, and as we analyze opportunities to rebuild infrastructures to accommodate more energy-efficient lifestyles we need to keep the long-term perspective in mind.

One basic problem of public transportation in an area with low population density will be the ability of people to access the transportation system. It may be that many people live several miles from the closest bus stop or station. The facilities for walking or riding a bicycle may not be very well developed. Maybe, the minimum requirement in such a situation will be access to convenient parking close to the station, so that people can use their private cars to get there. It may also be possible to restrict access by car, or to some types of car, such as ones run on fossil fuels, to some areas, and introduce mandatory "Park and Ride"[10] facilities in the future for people who drive fossil-fueled cars. It is also possible, but probably relatively expensive and resource-intensive, to administrate on a large scale, to encourage people to participate in large-scale carpooling.

The long-term development path for the development of public transportation and city planning could include the following steps:

1. Implementation of public transportation facilities on routes with a high number of travelers, where it may also be possible to offer "Park and Ride" solutions.
2. Increase population density, through the construction of new apartment buildings and offices within existing districts.
3. Increase shopping opportunities and other service facilities close to public transportation nodes, so that commuters can go shopping on their way to and from work and other activities.

CHAPTER 19

Transforming Heating and Utilities

The transformation of heating and utilities is different from the transformation of transportation. While the components of transportation systems, such as cars, trucks and fuels, are the same all over the world, the heating and utility systems differ between countries. In the United States, natural gas is used to a large extent for the production of electricity, making this country the largest user of natural gas in the world. In this country coal and nuclear energy are also used for the same purpose. In another large country, France, the natural endowments of gas, coal and oil are not very large, and this country has grown highly dependent on nuclear energy. In Sweden, with its large rivers in the north, hydroelectric power is complemented by nuclear energy to supply its small population of 9 million people with power.

While electricity has largely been produced in centralized power plants, and distributed through grids that encompass the whole of society, heating and cooling have largely been produced for single households or apartment houses at a time. Different power sources and systems have been used for this production. Coal and oil have been used, and in many countries households have turned to electricity for both heating and cooling, and more recently renewable alternatives such as geothermal energy, photovoltaics and pellet burners have been implemented on a larger scale in some countries. In many countries, power plants for the combustion of waste, or natural gas, are built that produce district heating, or heat in combination with electricity (CHP = combined heat and power, or "cogeneration") in large-scale facilities.

To make things even more complicated, many of the solutions that are available for large-scale heat and electricity production are also available as solutions for single households. While large-scale solutions may in many cases provide services at a lower cost of production, large-scale solutions may not always and in all situations represent the best alternatives. In situations where new infrastructures for the distribution of heating (or cooling) may be necessary, lower levels of investment will be necessary to install small units for single households or medium-sized units for a neighborhood.

19.1 THE ORGANIZATION OF A CHANGE PROGRAM

In this sector there is a need for volume transformation so that a larger share of renewable energy is produced by a certain date in the future. The total volume of renewable energy that is needed at different points has to be calculated, and a plan needs to take into account both the total need and the contribution to this production volume from different renewable technologies at each point in time. In case the construction of new energy production resources that are based on nonrenewable fuels need to be part of the mix, the investments in these technologies also need to be coordinated as part of the total project. In addition to these subprojects, a project concerned with the refurbishment and capacity increase of the installed base may become necessary. A suggestion for an organization structure for the transformation of heating and utilities is shown in figure 19.1.

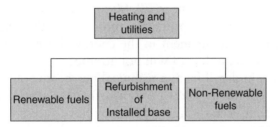

Figure 19.1   Suggested project organization for large-scale managed change project in heating and utilities

The activities in each subproject may differ widely. In the case of some technologies, there are already products and systems in existence that can be ordered and built. Maybe, the available amount of production resources need to increase and this may be part of the program, but otherwise it may be a matter of a rollout of existing technology. In other cases technology development may not be finished and this may have to be put on a "crash course" in order to make large-scale investments in the implementation of this technology possible.

19.2 TRANSFORMING PRODUCTION

As opposed to transportation there already exist a few alternative and renewable energy sources for heat and electricity production, and these alternatives are many times available as both large-scale solutions for large power plants, and as small-scale solutions for small houses. This could lay the foundation for a transformation process that to a certain extent could be market-based. However, the need to rapidly increase the capacity for the production of energy from renewable sources and the need to increase the overall production capacity for electricity in order to possibly facilitate the transition to electric vehicles add complexity to the situation. In addition to this, some solutions that are attractive short term, such as coal, may be limited for the longer term, because of expected peaks in the supply of raw materials. Apart from these issues related to the expansion of capacity, there is the problem of replacing existing nuclear plants and other power plants. Many of the existing nuclear plants are approaching the end of their technical lives and need soon to be replaced or refurbished.

In addition to this, there is the opportunity to build new plants for cogeneration or refurbish existing plants with this aim, which is a substantially more energy-efficient process than that of traditional power plants. While electricity production in a power plant only achieves some 30 percent energy utilization, the CHP process increases the utilization of the energy content to 70, 80 or as much as 90 percent.

Since this is not a book primarily about renewable energy sources, but about change management in the energy sector, and I am not an expert in energy technology, I will not attempt to try to analyze different alternatives in detail from a technical perspective. Instead, we will discuss here the main aspects of a transformation strategy. We have to keep in mind that even though we already have access to a number of renewable energy sources, the investment needs are so large that we need to identify the alternatives that will provide the largest improvements at the lowest cost.

19.2.1   Transforming Production by Improving Existing Power Plants

It is often possible to refurbish existing power plants in order to increase productivity, install cogeneration opportunities or facilitate operations based on renewable fuels. These opportunities need to be taken advantage of. When we discuss opportunities to use biological fuels or waste that could also be used for the production of bio-fuels, it is important that we calibrate strategies and plans between the two streams of transportation and power generation in order to avoid double count of the raw materials potential. We must also avoid building power and heat generation facilities to the point where they take advantage of too large amounts of scarce raw materials. We may need to use a share of our biological raw materials for transportation in the future, and we may be well advised to increase our reliance on wind and wave power, solar energy and other renewable alternatives for electricity production. Maybe we will be able to produce sufficient amounts of energy based on renewable raw materials in the future. The main issue may then be to increase the speed of growth of these sources in order to avoid energy shortfalls.

19.2.2 Transforming Production Using Renewable Energy Sources

There are already a number of new technologies available, which use renewable energy sources for heat and energy production. As a society, we need to define an optimal mix of such energy technologies, which could be used as a guideline for a large-scale transformation and implementation project. We already possess a number of technologies, which will be briefly discussed below, and a number of ideas for other technologies, that we could probably develop for large-scale implementation in the near future. The technologies that will be discussed are

- solar thermal energy and photovoltaics,
- wind power,
- bioenergy,
- hydroelectric power,
- wave energy,
- tidal power,
- geothermal energy, and
- nuclear power.

19.2.2.1 *Solar Thermal Energy and Photovoltaics*

The available solar energy on Earth amounts to a total of 10,000 times the current consumption of energy from fossil and nuclear fuels. Despite this abundance, a high share of the world's production of photovoltaic power until recently came from one plant in the Mojave Desert in California. In the case of solar thermal energy, the use is more widespread, and solar panels for the heating of houses and hot water tanks are increasingly installed on rooftops in regions with temperate to warm climates.

All forms of solar energy generation require the use of solar collectors, which require a certain area for their application. These collectors are becoming increasingly complex and efficient using mirrors to reflect and concentrate light for improved efficiency and reduced cost. Regardless of these improvements, the cost per unit of photovoltaic power is currently four times as expensive as that of wind power.

Because of the fact that land is expensive, the advantages of solar energy could be utilized to their maximum when otherwise unproductive land or areas in cities, such as rooftops, could be used. Thus, any house could be fitted with solar panels on the roof and even in temperate climates this could contribute substantially to the heating of water for a household. A system made up of a collector panel of between 3 and 5 square meters, a storage tank and a pump could typically be expected to supply between 40 and 50 percent of the hot water needs of a typical European or US household.

Solar thermal power, collected through rooftop collectors, could be used extensively for the purposes of heating water and houses. Due to the need of daylight, and, in the case of the advanced solutions using mirrors, direct sunlight, this source is intermittent and can only be used for production during hours of daylight. The main use of photovoltaics will thus be as an additional source during daytime peak hours. For solar energy to be used at other times during the day, large-scale development and production of batteries for storage would be required. This source of power could possibly be profitably used in the future on a small scale for individual households and small neighborhoods and on a large scale in deserts and other areas with a high level of direct sun radiation and an abundance of cheap and unused land.

The high cost of photovoltaics may partly be due to the low demand for photovoltaic panels, and the production cost may change rapidly in case demand increases or if government subsidies change the cost equation. Subsidies of this technology that could lead to large-scale investments in photovoltaic power plants may be necessary to push this development forward. It may be estimated that the development of photovoltaics is at the same level today that wind power was 20 years ago. It has taken that much time to take wind power from the experimental stage to the present situation at the threshold of large-scale expansion, at which it now supplies less than 1 percent of the electricity needs in the United States and Europe.

19.2.2.2 Wind Power

Wind turbine technologies are, from the generation cost perspective, competitive with the cost of power generation from nuclear and fossil-fueled plants. Wind power is also already being used all over the world, and a number of large-scale wind turbine projects are already being planned for construction over the next decade. Similar to photovoltaics, wind only provides intermittent energy and these sources need to be complemented by more reliable alternatives and not all locations are ideal for wind power generation.

The need to increase the power production from renewable energy sources call for a large-scale increase in this capacity. Wind power is currently the best alternative for this development. In order to fulfill plans for wind power to supply 12 percent of the global electricity needs by 2020, as suggested by the European Wind Power Association, 630,000 new 2-megawatt turbines need to be installed.[1] This requires massive expansion of the current capacity for wind turbine production. This capacity amounted to 20 gigawatts per year of new capacity globally in 2007. In an energy transition program, the opportunities to expand power generation, using wind turbines, need to be explored, and the capacity to fulfill the goals set, or possibly to exceed them, needs to be put in place. This will require massive investments in a number of areas, including production capacity for wind turbines, investments in the wind turbines themselves and in transmission capacity from the wind farms to customers.

19.2.2.3 Bioenergy

Bioenergy is a term that is used collectively for energy from different types of biological materials, including biological waste, which may be burned directly, or turned into biogas, biodiesel or synthetic diesel. The arguments regarding biofuels in the transportation section are to a large extent valid for this section as well. The use of bioenergy as a source for the production of heat and electricity competes for raw materials with the use of bioenergy as a fuel for transportation and the availability of raw materials and land for the growing of energy crops will represent important limiting factors for the expansion of energy systems that are based on bioenergy.

19.2.2.4 Hydroelectric Power

Hydroelectric power has been used for centuries in order to harness the power of rivers. In some countries, such as Sweden, the potential for the expansion of hydroelectric power generating capacity has almost been used to its full, but in other countries there are still opportunities to use this resource at a small or large scale. The production technology is mature, production cost is low, and the main obstacles are environmental concerns regarding further capacity expansion. There is also the time aspect of the large-scale projects that is necessary. It takes a decade from start to finish of a large project.

19.2.2.5 Tidal Power

Tidal power could be made into an abundant and reliable energy source around the coasts of many countries. In the same way, as in the case of hydroelectric power, the technology for the production of tidal power is based on containing large volumes of water that rises through the tide, in large basins, in order to release the water through turbines and, by doing this, generate electricity. The tide is caused by the regular movements of the Moon in relation to the Earth, which makes this source of energy more reliable and predictable than wind and wave energy.

At present, tidal energy provides very small amounts of energy globally, mainly through a French project at the Rance estuary.[2] It has been estimated that one possible site at the Severn estuary in the UK could provide almost 5 percent of the total electricity consumed in the UK in 2002.[3] There are a number of different sites in the UK and other countries that could be relevant for this type of development, but the environmental concerns related to these developments are also substantial.

Considering the large-scale nature of this type of project, we will probably have to wait for more than a decade from the decision to go forward, until this type of installation starts to generate electricity. From the perspective of the transformation program, we need to make quick decisions about this type of project and include the development in the long-term or medium-term plans for renewable energy sources. Due to the size of projects, large amounts of tidal energy will not become available within the next decade.

19.2.2.6   Wave Energy

It has been estimated that the globally available energy resource from wave power amounts to 17,500 terrawatt-hours, which is more than 1000 times the electricity that is currently used globally.[4] To many countries, including the United States and the UK, the exploitation of this resource is a very promising prospect.

However, there are as yet no full-scale installations of wave power plants, and the technologies under development are based on a number of different types of mechanisms. There are ideas that are similar to the technologies that are used for hydroelectric power and tidal power. In these cases water that is set in motion by waves is captured in a reservoir, and released through a turbine at will. In this way, the energy is stored in the reservoir until it is planned to be used, which increases reliability. Other ideas focus on capturing the energy of the constantly moving water at sea, or along the shores, through propellers, or "buoys" connected to power generators.

The operating and maintenance cost of power generation, using wave energy solutions, is reduced with each new generation of technologies that is developed. It is probable that some of the solutions that are being developed will become economically viable in the future, but the experimental stage of these technologies indicates that we are looking at a 20-year time frame for the commercialization. This is based on the time that it has taken to develop wind power from a similar stage to the present early commercialization stage. The existence of a large number of ideas that are currently being funded in parallel through private and, to some extent, public funding will probably increase the time until a viable technology gets ready. In case we want to speed up development, a small number of technologies should be selected, each being the most promising for a particular area of generation. The funding of the development of these technologies should then be increased to meet the needs of a high-speed development project. In case time is not seen as a constraint, development could continue along its current path. Critical factors to reduce electricity production cost, and probably also the cost of production and maintenance of the units themselves, are the number of moving parts. These need to be kept to a minimum in order to achieve a high level of robustness to survive the harsh sea environment for decades. This creates an advantage for dam projects similar to those for tidal energy, with the environmental downside of such large-scale constructions.

It is clear that solutions, which include a number of turbines of 1 or 2 megawatts each, are completely different, in terms of the project components, compared to that of building large-scale tidal or wave power dams. The size of each project will be smaller, more of the size of wind turbines, and a large number of installations will be necessary in order to generate large amounts of electricity. Thus, the problem of implementing wave energy on a large scale will also be that of building production resources for the manufacturing of large numbers of wave energy generators.

Thus, we have two key lead times to reckon with in the project. The first is the time it will take to develop a wave power technology that can be used for series production of generators. The second is the time it will take to increase the production capacity and installation capacity for these generators to the volumes that will be needed. Most probably, we need to wait another decade, or more, until we will see wave power generators produce even 1 percent of the total energy needs of any large country. What we can do right now, however, is to make an estimate of the lead times under different conditions that could form the basis for a decision on how to run the project. We can also read the book by Professor Ruttan in which he explains the role of government in the development of general purpose technologies, and ponder the implications of this for the transformation project.

19.2.2.7 Geothermal Energy

Geothermal energy is derived from the heat that is generated inside the earth, and which can be converted to thermal energy or electricity. Geothermal energy is available in many forms, in hot springs that are located in a few places around the world, but also in the form of ubiquitously available heat in the ground, that could be captured by ground source heat pumps. Energy from hot springs and volcanic energy (high-enthalpy resources) can be used both for power and heat generation, while the ubiquitously available low-level energy (low enthalpy) can be used for heating only. In this section, we will focus on the low-enthalpy[5] applications of heat pumps, because they represent an alternative that is available everywhere and at low cost.

Heat pumps for small- to medium-scale heating needs (single houses to city districts) are available, and they have been used for purposes of heating houses. Heat pumps are powered by electricity, and they produce approximately three or four times the amount of heat compared to the energy needed to power the pump. With increasing energy prices, these pumps pay back the investment made in them within a few years.

Box 19.1 Geothermal Energy at IKEA

IKEA globally runs 270 stores and 30 distribution centers. The goal is to reduce energy consumption in its global operations by 25 percent, within the next few years. Another goal is that of changing to 100 percent renewable energy sources for electricity and heat within the same period of time.

The latest addition to its network of stores in Sweden, which is the country of origin of IKEA, is the store in Karlstad, which uses geothermal heat that is extracted from subterranean rocks for heating. In addition to this new store, three old stores will be replaced by new buildings. The two new stores in Uppsala and Helsingborg will be heated through the same technology as that in Karlstad, and the Malmö store will use heat from subterranean aquifers.

In order to achieve the goals for renewable energy and energy reduction, each IKEA store is encouraged to identify locally available opportunities to use surplus energy or renewable energy sources. The advantages of geothermal energy are many, including the fact that this source of renewable energy for heating will not run out, and it is available everywhere. The investments are small enough to pay back in only a few years.

A large part of the growth in the installation of geothermal energy units has been in the form of heat pumps for single households.[6] These installations extract heat from the ground through piping, which has to be dug into the ground or drilled into the bedrock. The cost of purchasing and installing this type of unit will amount to some 10,000 dollars for a single household. Larger installations for multiple households could be achieved at a lower cost per household.

Another type of heat pump extracts heat from air, through a reverse refrigeration process. This type of pump is less expensive to purchase and install, and it delivers heat in the winter as well as in the summer, also in cold climates.

For the purpose of the energy transition program, the various types of heat pumps represent viable small-scale heat-generating units, which are affordable for single households. The energy saving is substantial, and the payback time may be 3–10 years, depending on the size of pump used and the climate conditions. There is already substantial production capacity in place for these units, the financing of such investments is simple, and neither the installation process nor the maintenance requires large amounts of time or other resources. In combination with an increasing amount of electricity from renewable energy sources, this represents a very attractive alternative for the short term.

19.2.2.8 Nuclear Power

There are many ways to look upon nuclear energy. One aspect is the resistance to the technology among many people, because of accidents or near-accidents in the past. Another aspect is the problem and cost of the final storage of used uranium and plutonium. The aspect that speaks in favor of nuclear power is the high energy content of small amounts of radioactive uranium. This energy content could be even better utilized in the new fast breeder reactors, than in older reactor types, and the cost of production could, potentially, decrease. At the same time the sources of uranium, using the new reactors, could last several hundred years at the current rate of production and more than a century if we increase our production of nuclear energy. We may even be able to reuse uranium and plutonium that has been used in the old and less efficient types of reactor.

From the point of view of the energy transformation program, we have to focus on the fact that nuclear energy provides carbon dioxide–free energy. The time that it takes to build new reactors is today, according to industry experts, less than 10 years and several reactors are already under way globally. Even if the steering committee of a program, perhaps headed by a prominent official in each country, would set the transformation program on a crash course, and press for a decision to embark on a large-scale nuclear energy expansion program, as is currently discussed between the UK and France, we would have to wait for, perhaps, 10–15 years until we get the first kilowatt-hours from new nuclear plants in this program.

Investments in new generation facilities for nuclear power must be seriously considered in many countries and in some countries new projects are already under way. The technologies used in modern plants have been improved compared to the installed base, in terms of capacity, utilization of raw materials, cost and safety. The safety issues are still being debated by industry experts and environmentalists. Part of the picture framing this debate, however, is new. Over the next decades we need to build enough new clean energy resources to replace many existing plants and probably to increase the capacity of power generation in order to replace fossil fuels for transportation and coal-fired plants for electricity generation. In addition to this we need to either increase production capacity compared to current supply or make large structural investments in order to reduce electricity demand. None of these alternatives are for free.

Reducing energy demand may, for all we know, cost as much as increasing capacity, but a reduction will nevertheless be better for the long term. In the short term we will probably need to do a little of both. What is on the line, and this we need to keep in mind, is the ability of our society to support the lives of a growing population of upward of 6 billion individuals on this planet. The production and distribution systems in our society are geared toward high energy demand for all kinds of purposes that are necessary for our well being and our ability to earn money and this could not easily be changed. Unrealistic claims implying that energy demand could be reduced at almost no cost need to be substantiated almost immediately or discarded. For the time being we need to expect that this is not the case and make cost estimates for both reduction and expansion alternatives. In all probability, we will need a large amount of new nuclear capacity in the near future and we are likely to find that the alternatives to this expansion are more threatening than the nuclear power plants themselves.

19.2.3   Many Things Need to Be Done at the Same Time

In the way that has been indicated above, there is no single solution that can be used to solve all of our energy problems in one go. In the same way that we will need to develop a number of engine technologies and renewable fuels in parallel for the transformation of the transportation sector, we will need to apply a number of different technologies for the generation of heat and electricity. We will need to combine both the different energy sources and generation technologies above, and we will also need to apply both small-scale and large-scale solutions.

Overall, for the transformation within the program stream of "Heating and Utilities," we need a strategy for the extent to which we will use each of the different alternatives, and a plan for the stream as a whole, and for each project within the stream.

Below, some aspects related to large-scale and small-scale alternatives will be discussed.

19.2.3.1   Large-Scale Generation

The advantage of large-scale generation alternatives is that a smaller number of large-scale projects could generate substantial amounts of renewable energy. Large-scale projects can be managed as large projects, which we have substantial experience of running, and each of the projects concerns the large-scale implementation of technologies. Within the large-scale alternatives we find that different choices will create different production structures in the future. In the case of nuclear energy and tidal power, a smaller number of facilities will be needed than in the cases of wind turbines and wave energy. These alternatives represent completely different types of technical and financial challenges and they may all be needed to some extent in the future structure for energy production. Compared to small-scale generation for single households or apartment buildings, the large-scale technologies require fewer resources per volume of production to sell, construct, build and operate, than it does to construct, produce, sell, deliver and operate a large number of small units. In a situation where we have a large energy transformation program in society, covering a number of streams, which requires resources in all streams for a number of purposes, it will be an advantage if we are able to achieve a substantial effect using limited resources. It makes sense to initially exploit the opportunity to install a substantial number of large power plants and generators, using different energy sources and technologies in densely populated areas. We may also apply small-scale tried and tested technologies, such as heat pumps, in areas where these are relevant.

We have access to a number of different technologies, and there are production resources available for a number of different large-scale alternatives, such as cogeneration power plants, wave and tidal power generators and wind power generators. These different alternatives show different cost profiles for the production of plants and generators for the same amount of energy. These differences need to be taken into account in the strategy formulation phase of the program.

In the case of wind, wave and solar energy, energy is abundant, but intermittent, and there is virtually no limit to the volume of energy that can be generated, using these technologies, but the cost of land and the cost of constructing a large number of photovoltaic cells may preclude the establishment of large-scale photovoltaic power complexes in areas with a high cost of land and the cost of electricity production using these technologies will for a long time be higher than the cost of wind power production. In both cases, generation has to be complemented by other more reliable sources. Thus, in each case, there are pros and cons, and the different alternatives need to be evaluated before a detailed plan for the longer term is developed. In the short term, however, we can already see that we will need a number of units of each of the alternatives that is already viable, and we could relatively rapidly develop a rough short-term plan that includes investments in wind power and cogeneration. Wave energy and photovoltaics may be possible to finalize and implement on a large scale in a decade or sooner if development is set on a crash course. Other large-scale alternatives, and the detailed goals for each of the given alternatives, will need more elaboration and analysis.

19.2.3.2　Small-Scale Generation

Small-scale generation alternatives have many merits, and some of them are well tried and available at reasonable cost in large quantities. The increased application of heat pumps, solar thermal collectors and boilers for the combustion of biogas or solid fuels, such as pellets, are probably given, for some customer segments and regions, while other segments and regions may still be better served through large-scale applications.

CHAPTER 20

Transforming Industrial Processes

In the United States, 36 percent of all energy is consumed in industrial processes and another 16 percent is consumed in commercial buildings. There is an opportunity to more than halve this consumption, using existing technology. This will be an important step, which will also be relatively easy to achieve. It will be necessary in order to reduce the need for new electricity-generating capacity. One of the main activities that will be needed in order to achieve this transformation will be a large-scale training and information program, to spread information about savings opportunities, actions and financing opportunities to companies of all kinds.

Companies in different industries and with different types of businesses have very different opportunities to save energy in their production, purchasing and distribution systems. In the first place, the amount of energy that is used, and thus the amount which can be saved, differs widely between industries. There are industries, which are highly energy-intensive, such as heavy industry. There are also industries that are less energy-intensive, but where companies, nevertheless, have large energy savings potentials. This goes for most large companies that in different ways use large amounts of energy for production, heating, lighting and distribution. Since the number of companies with less energy-intensive businesses is higher than energy-intensive ones, it is important, within the industrial systems stream, to analyze savings opportunities in all kinds of companies, and also develop knowledge about particular savings opportunities in different types of companies.

Besides the savings opportunities that exist in key business processes, there are also savings opportunities in support processes. These are more similar between companies in different industries. Often substantial savings opportunities with rapid payback cycles exist in the fields of lighting, heating and ventilation and pneumatics.

In this case we need to keep in mind that savings opportunities could be realized more rapidly, in a small number of large, energy-intensive companies, than in a large number of smaller and less energy-intensive companies with large differences among them. However, it may be difficult, or downright unfair and possibly damaging to the competitive position of these companies and also to economic growth, to place the entire burden of energy transformation on energy-intensive companies. On the other hand, the complexity and cost of running a broad energy savings program across all of industry will be high. Yet, this may be what needs to be done. In most cases, the projects will be able to make use of existing technology and consist of a large number of different activities, each of short duration. The complexity and novelty of administration of the program in itself will be high and both training and incentives will need to be used in order to achieve optimistic goals.

Further, on the positive side, however, projects in this stream will prove to be less complex than those related to the transformation of transportation. This is because substantial savings could be achieved through the large-scale implementation of existing energy efficient technologies and known process improvements in companies. In industry, energy savings have to be counted by the different sources of energy that are used in order to fuel the industry of any country and the calculation has to take into account both the savings in each type of energy and the increased need for another, in case switches from one energy source to an alternative is made.

According to *The Guide to Energy Management* by Capehart, Turner and Kennedy, companies could often reduce energy consumption by 70 percent in office buildings and halve it for manufacturing plants. Some 60–70 percent of savings could be achieved within 3 years from the start of a program in any particular company. However, with the large number of companies that need to go through a savings program, this type of program will require more than a decade for completion, unless it is put on a crash course. At current energy prices these savings make financial sense. While the figures represent the savings potential in individual companies, a program covering all of industry would take longer. However, considering the small amount of time it took during the Second World War to transform the entire American industry to war production, the low technical complexity and the financially sound basis of many of these savings, this part of the transformation program may be the easiest and quickest to realize. One of the main issues will be the expansion of production capacity for the necessary new equipment and consumables.

Box 20.1 Modern Motors from ABB Have Saved 14 Nuclear Power Plants

The sale of modern frequency-controlled motors by ABB has saved the energy equivalent of 14 nuclear power plants. Still only 20 percent of all motors in the world are frequency-controlled. All in all, it has been estimated that more than 50 percent of the total use of electricity is used to power electric motors.

A substantial amount of savings could be made on electricity. These savings are included in the above figures and they will be important to take into account in relation to the transformation of transportation and electricity production and generation. If transportation is transformed, so that an increasing number of vehicles will be charged from the electricity grid, this will increase demand for electricity. In order to reduce the need for substantial investments in new power-generating capacity, and be able to transform existing power generation to more sustainable alternatives, we will need to reduce overall demand for electricity, and not increase the demand more than necessary by moving consumption of other fuels over to electricity.

20.1   THE ORGANIZATION OF A CHANGE PROGRAM

A change program in industry will need to contain a number of different subprojects, including production and support processes, building installations, transformation to renewable energy sources, logistics and entrepreneurship and new business development. Within the production and support process subproject competence in different energy technologies needs to be built, so that support could be given to companies in different industries (see Figure 20.1).

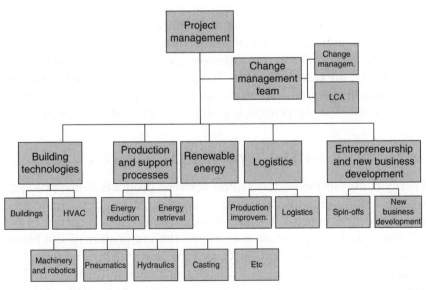

Figure 20.1   Suggested project organization for large-scale managed change project in the field of industrial processes

20.2 ANALYSIS OF ACTUAL SAVINGS OPPORTUNITIES IN COMPANIES

In the diagram above we can see how energy consumption breaks down into a number of broad areas of consumption. Within each of these areas in one particular company, there are dozens, and in large companies hundreds, of individual improvements that could be made in order to improve energy efficiency. A project in any one company needs to start with an analysis of energy consumption in different areas, and identify opportunities and estimate the investment need and savings potential of each of the opportunities. Based on this, each company needs to make a plan for the transformation and start to realize them in a dedicated program.

Apart from the overall energy use, electricity use could (according to the US Department of Energy) be broken down into pumps, which represent almost a quarter of all energy use; nonmotor use that stands for almost as much; fans, blowers and compressors, that consume a similar amount; and machine tools, other motors, DC drives and HVAC which make up the rest. The analysis needs to be based on an understanding of actual energy consumption in each company, rather than on blanket figures. Energy managers need to analyze patterns of equipment use, energy bills and technology development in the different areas of relevance to each company. There are a number of books and publicly available energy management methods to support this endeavor. There are also consulting firms that offer support and management of these efforts. The number of people who work with these tasks will need to increase in the near future.

The most important part of any change program is management commitment. In order to transform energy use in the entire industry of any country, the commitment to change within each company will not be enough. An overall change program run by government will probably be needed, if the process is going to proceed at high speed.

20.3 PRODUCT LIFE CYCLE ANALYSIS (LCA)

Energy consumption patterns differ between products. Some products, such as aluminum and paper pulp, require large amounts of energy in the production process. Other products require much energy in distribution, and many products consume the major part of the energy during the lifetime of the products, as is the case with production equipment in industry, cars, freezers, washing machines and other household appliances. A number of products, such as batteries, require substantial amounts of energy in the destruction and recycling processes.

In an effort to reduce energy consumption, an analysis needs to be made of energy consumption in the different phases of the product life cycle. In this way, energy savings opportunities can be identified in different parts of the life cycle and opportunities for energy savings can be identified, which were not identified in the general analysis of the business above. This analysis may render the result that a product should be constructed in a new way, in order to save energy when the product is used, or that the destruction process could be run in a different way, in order to reduce energy consumption in this part of the cycle.

These analyses may sometimes render savings opportunities that could be achieved through the use of existing technology and in a relatively short time. The pursuit of some opportunities may require new technology development which takes time and costs money.

Box 20.2 Attention to Energy Consumption through the Product Life Cycle at IKEA

IKEA has always focused substantial amounts of effort on the reduction of resources used in its products. IKEA sees it as its mission to improve the daily lives of the many, and this translates into making affordable products. In order to make products affordable, the company, based on the cost-conscious culture that was firmly implemented by its founder Ingvar Kamprad, has made resource efficiency into a virtue.

With meticulous precision, designers at IKEA use resource efficiency as an important constraint in the design of products. Products are designed to use as little wood and other materials as possible, and methods and technical solutions for assembling furniture are constantly improved by designers. Designers use as few as possible of the standardized screws, nuts and other implements that are necessary in order for the customer to assemble the product. Designers even adapt the measures of products, and their mostly flat packages, to the sizes of standardized transportation pallets. In this way the company ensures that the available loading space of a typical freight container, or truck, is used as efficiently as possible.

Some of the 10,000 products of the existing product line are even redesigned for improved energy and cost efficiency. One of the most popular items at IKEA is the couch Klippan. This product has been sold by IKEA for almost three decades and it has sold several hundred thousand units. When it was first launched, it was preassembled at the factory, which means that it was delivered as a finished product to the customer. In this way, however, a pallet could only hold two units of the product. Applying the logic that permeates the whole organization, the product manager for couches, together with a designer, decided that the product should be delivered unassembled to customers. By doing this, each pallet could hold four units, decreasing the transportation and storage space for the product by 50 percent. In return for the time that the customer now spends in assembling the couch, they get the product at a lower price, achieved both through reduced cost in assembly and through the improved use of transportation and storage space.

This is one example of a company that achieves cost leadership in its industry through the application of an almost single-minded attention to resource efficiency. From the design of the product, through the choice of materials, to production, distribution and sales, IKEA pays meticulous attention to every detail. Most importantly, IKEA never compromises on quality. Through extensive testing of every aspect of use, the durability and quality of every product is tried according to the most demanding standards of the industry. The same is true for the appearance of products, which is another overarching goal for the designers. In order for a product to become successful in the IKEA product line, it not only has to be resource-efficient and durable, but the visual aspects of its design must also appeal to the most demanding of consumers. The testing procedures and other attention to quality at IKEA ensure the durability of products, so that customers do not need to replace furniture as frequently as with products of lower quality. This durability, too, contributes to the overall energy efficiency.

20.4   RETURN LOGISTICS AND THE REVERSE FACTORY

One idea that may be developed in the future is the idea to return used products to a reverse factory, at which they are disassembled and the parts are reused in new products. In order for this to become possible, products need to be designed for disassembly and parts need to be fastened using materials that make it easy to break them down into parts again. Xerox has already successfully tried this for copiers. A large number of parts turned out to be reusable, even in older generations of copiers that were not designed for disassembly. The Italian automaker Fiat has developed a concept car using these principles. This car was disassembled by robots in a factory and the parts were sorted into containers for reuse.

This type of return logistics system could be seen as a new way of creating value and when it has been fully developed into a smooth and cost-efficient system, it is expected that it will save both energy and money compared to the present situation.

20.5   TRANSFORMING PRODUCTION TECHNOLOGIES AND CHOICE OF MATERIALS

Gradually, since the Industrial Revolution, the energy efficiency of production technologies and materials has improved. Even if energy efficiency has not been an overall priority, improved energy efficiency has been a side effect of the development of better production technologies. The improved efficiency has been achieved through many different development trends. One of them is, naturally, the development of increasingly efficient motors for machines, but more importantly the improved efficiency of machines is due to the increasing speed of machinery and production processes, which allows a particular step in the production process to be performed more rapidly.

In the same way, lighter and more advanced construction solutions have reduced the energy content of products. One example of this is the development of lighter airplane frames during the Second World War, which made it possible to build lighter aircraft that could fly longer distances. The trend toward lighter and stronger constructions has continued product generation after product generation. Developments of new materials has added strength to products, and at the same time reduced weight. The replacement of steel and wood by plastics is an example of this trend, which reduces the energy content of products despite the fact that plastics are made from oil. In case we would be forced to go back to metal, wood and other heavier materials, which require more energy in order to be formed, we would lose energy in more than one way. First, we would need to use more energy in production, and second, we would need to spend more energy in transportation, moving heavier and bulkier goods around from production to consumers.

This development of production technologies and materials needs to continue, and we need to continue the development toward the use of an increasing amount of IT, which makes it possible to further reduce the energy consumption in production. One of the consequences of the use of IT in industrial equipment is that one piece of machinery could now be programmed to perform several production steps in a sequence, without transportation between different machines and other additional, energy-consuming activities in between.

We will here, as examples, describe two different technology developments that will contribute substantially to energy efficiency, as they gradually replace existing technologies.

20.5.1   Digital Printing

Offset printing is gradually replaced by digital printing, which in many ways improves the energy efficiency of printing and publishing processes. Offset printing is a process that requires a number of different steps and materials for the production of a printed product. Involved in the process will be a printing machine, a cutting machine, a sorting machine and a binding machine. The printing machine in itself is a heavy piece, weighing several tons, consisting largely of steel and heavy mechanics. The printing process involves a number of steps for the printing of each book, magazine or brochure. A set of pages are printed on a sheet, in the number of copies that are needed in the run. After all sheets have been printed, they are cut into the right format and the pages are put in a sorting machine and sorted into the right order. The cost of setting up a production run is so high that it is better to print 20 percent extra copies, than to be forced to print an extra run if the product runs out. Therefore, most brochures and other printed material are printed in excess volumes, in order to avoid reprints at high cost. These excess volumes represent waste both in terms of energy and money.

Digital printing, on the other hand, works in the same way as an office printer. All the pages of one book are printed in the right order. They do not need to be sorted, and the final product could be bound or stapled together immediately as each copy has finished printing. The cost of printing a copy of a long production run is slightly higher than in offset printing, but the higher cost is often offset by the reduction of other costs, such as the cost of setting up a production run, which is very low. Companies that use digital printing can afford to print only the number of copies that are needed, and production runs could be made shorter, because it is possible to reprint without an exorbitant extra cost. This allows for frequent changes in catalogs during a year, whereas previously, the new products had to wait to enter the catalog until next year's printing.

Using digital printing, energy consumption is reduced through the need to only print the number of copies that are necessary at each time. The printing of each copy is also less energy-consuming, with fewer heavy parts to move around inside the printing machine.

Another source of future energy savings will consist of the opportunity to break large print runs up into smaller runs that will be printed in the same region where it is going to be distributed. In this way, for a company that is going to print a brochure for distribution across a large country, the need for each urban area could be printed by a print shop in the region in question. In this case, the transportation of the final product would be reduced to the distribution within each region. The opportunity to print the same product in different places could be offered by a company running multiple printing offices in a national network, or by cooperating printing companies, offering the same service through cooperation.

The complete departure from printed matter to digital distribution of information would naturally reduce the energy consumption even further, and this might be the next step in the development, or a step in a parallel development, depending on the situation.

20.5.2   Plastics Used for Automobile Bodies

Car bodies are mostly made of steel and aluminum. For a long time, however, automotive companies have researched the use of plastics for car bodies. The problem, so far, has been that plastic cars have been too inflexible, and thus too dangerous and damaging for metal cars in a crash. Nevertheless, the advantages of plastic car bodies, in terms of energy efficiency, would be tremendous.

The molding of plastics is a process that is more energy efficient than the production of steel and rolling it into rolls of steel plate. The transportation of plastic granules is also less energy consuming than the transportation of rolls of steel to automotive plants and the lighter plastic car bodies will help in the reduction in the fuel consumption of cars. In the production of a body, the molding of a plastic car only requires one production step, which of course requires some heat, in order to melt the plastic granules and mold them into a product. Compared to operating a hydraulic press, that weighs several tons, together with the production of the steel, the energy saving will be substantial. The plastic body does not need painting, because the plastic granules already are of the right color and the transportation between steps in the process from steel production to final assembly will be reduced as well.

In total, most plastic products are more energy-efficient than their metal counterparts, and this will be true for the plastic automobile as well.

20.5.3   Low-Value Plastic Products

Plastic car bodies and plastic parts for machinery are examples of high-value applications of plastics. If plastics in these applications were replaced by metal, the energy bill would increase. There are also numerous low-value applications for plastics, such as for packaging materials and cheap toys. In some cases it would be easy to replace the plastic by other materials, such as in the case of the replacement of plastic shopping bags for paper. In other cases it may be less easy to find alternatives and the alternatives may demand lifestyle changes, large-scale changes in production and distribution systems or reduced consumption.

In the case of flexible packaging for frozen foods or coffee these packaging materials have been developed in order to preserve particular food products in an optimal way. Flexible packaging materials consist of multiple layers of different plastics and, sometimes, aluminum. For a particular food, the combination of layers is selected because of the barriers that are needed for the food in question. Some of the layers are applied through the extrusion of very thin coatings of a particular material, in order to save material content and weight. For many foodstuffs the use of carton packages would reduce the storage time and increase the perishability of products. If we were to abandon plastics, we would need new production and distribution systems for some types of products. We may want to purchase smaller packages or turn to distribution systems that differ in other ways from the ones that we have become accustomed to. In the case of replacing frozen foods by chilled or fresh distribution, this would require large-scale changes in consumption patterns and logistics systems, since many foodstuffs would be available only in season and often in limited volumes.

In the case of plastic toys and other low-value products, people may increasingly turn to wooden alternatives or other renewable materials. This would require new production systems and people would afford fewer products. With increasing oil prices the price of plastics will increase in any case, but the consequences of such a development will have to be planned for. A reduction in the use of plastics may not be a bad thing in itself, but a reduction in consumption would also imply different effects on economic growth, employment and other important aspects of our economy.

We need to once more remind ourselves that we have increasingly turned to plastics, increasing global production and distribution networks and other changes in consumption patterns for a reason. The main reason has been convenience, increased consumption and economic growth. If we want to reduce our need of plastics on a large scale this cannot be achieved without repercussions through other parts of the economic, production and distribution systems as well. In the energy transformation program this may be necessary, but it will not be easy and it will not be achieved without strategic planning, information and program management.

20.6   DIGITAL SERVICES THAT REPLACE PRODUCTS

In the case of information services we can now replace existing physical products with electronic services. We have already described how a digital printing machine replaces a number of previous machines, which were made from heavier materials, and that required more energy both to produce and to run in production. In addition to this, offset printing has made buyers of printing services order more copies of each publication than are likely to be used. This situation is now slowly changed through the application of digital printing.

We have also, briefly, considered the possibility of replacing written material by digital products. Books, brochures and other printed matter belong to this category. Using a device for reading digital books and magazines, readers could keep whole libraries in electronic form in the future, which also will cut down on the need for book shelves and other furniture for the storage of books, brochures and documents. It would also reduce the need of storage facilities for books at publishers, transportation resources and room in distribution centers for these products. We would need less space in the form of bookstores, in the same way that we now experience how the volumes of records distributed in the form of physical CDs and DVDs are reduced as people get increasing amounts of music electronically.

In the same way we find that the opportunity to electronically receive invoices and tax statements from the IRS reduces the need for postal distribution of letters containing such messages.

Box 20.3 The Energy Advantages of Electronic Communication Solutions

According to a study financed by the WWF and ETNO,[1] savings can be achieved if electronic communication replaced traditional solutions:

- If video conferences were allowed to replace 20 percent of European business travel, this would reduce petroleum consumption and CO_2 emissions would decrease by 22 million tons annually.
- If 50 percent of all EU employees would replace one business trip by a conference call, this would similarly reduce CO_2 emissions by 2.1 million tons per year.
- If 20 percent of households in EU-15 countries would use a virtual answering machine instead of purchasing a physical one, this would save 1 million tons of CO_2 emissions and reduce electricity

consumption and reduce the need of oil for the transportation of the machines.

- If 20 percent of all EU companies that send out bills to more than 100,000 customers would offer them to pay via the Internet or via their mobile phones, this might save 0.5 million tons of CO_2 emissions.

The opportunity to replace physical information products by electronic alternatives on a large scale offers us the opportunity to reduce energy consumption in a number of ways. This type of change, however desirable, will also cause substantial change in economic systems, which needs to be planned for. However, the energy transformation will require a large number of new companies to be created and new products to be produced and distributed. This will require large numbers of employees in the resulting businesses. These resources will to a large extent need to be freed from existing companies, a process that may have to be facilitated through planning.

CHAPTER 21

Transforming the Built Environment

One sector in society that accounts for a substantial amount of energy consumption is the housing sector. Housing is closely related to city planning, through the need of each house to be connected to water, electricity, sewage and other systems. The population of a city needs access to transportation, schools, shopping and other necessities. In this way, city planners, and the legislators who set the rules for city planning, influence the long-term sustainability, or lack of sustainability, of a house, a district and a whole city.

Houses, and cities, are built for the longer term, on the expectation that houses will be used for 100 years, or more, and that cities will prevail, almost indefinitely, into the future. The best time to plan for energy-efficient cities is before most of the houses are built. It is during the construction of a new house that houses can be made sustainable at the lowest cost and it is at the initial stage of planning of a city that the plans for the networks of roads, railways, footpaths and other aspects that determine a large portion of the energy consumption are laid down. We will now go through three areas that are necessary to tackle within the stream of the built environment of the energy transformation program. These areas are

1. construction of new houses,
2. refurbishment of existing houses, and
3. sustainable city planning.

In the same way as earlier, we will not go into technical detail, but rather discuss the overall opportunities and the measures that need to be analyzed and possibly included into a strategy and a plan for the transformation within this stream.

21.1   THE ORGANIZATION OF A CHANGE PROGRAM

In the case of the built environment there are four key subprograms that need to be coordinated, as shown in figure 21.1. One has to do with building regulations and regulations for city planning. These need to be changed so that practices in the fields of construction and city planning may change. The second key subproject is concerned with

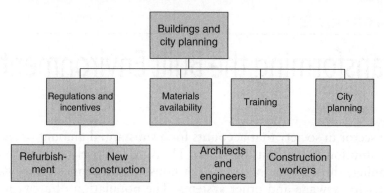

Figure 21.1 Suggested project organization for large-scale managed change project in the area of the built environment

securing the availability of materials and components for the houses that we need in the future, and for rebuilding the houses and communities that we already have. The third is concerned with the training of architects, engineers, construction workers and other people who have an impact on the way we build our houses. The last subproject needs to develop the principles and practices of city planners, so that they can contribute in an optimal way to energy efficiency.

At this point the exact content of this program is less clear than the components of the industrial processes program. We know of a number of possible solutions for construction in the future, and we know some things about how to implement them on a large scale in a short period of time, but we do not possess the whole picture. The development of the details of this program must be part of the planning effort that needs to be taken care of in the near future.

21.2   CONSTRUCTION OF NEW HOUSES

Houses represent a type of stand-alone systems in our categorization, which was presented earlier in this book. This means that a new house could be built using the most modern and energy-efficient technologies available, without regard to how other houses in the surrounding area have been constructed. For the purposes of the energy transition program, it makes sense to immediately implement as much of the experiences from recent construction projects for energy-efficient houses, in as many new construction projects as possible.

One type of energy-efficient house is called "passive house." Passive houses require a minimum of additional energy for the purposes of heating and cooling. In all parts of the construction, opportunities to recover and save energy have been taken advantage of. This reduces the need to heat the building, heat water and the need to have active ventilation systems. There are discussions within the EU to pass new legislation for the construction of houses according to passive house standards. Such a law would require large-scale change in the building materials industry as well as the construction industry. Due to the low turnover of houses, such a law would change the energy efficiency of houses very slowly. It may not have a substantial effect on the average energy consumption in more than a decade. In other areas of the world, where new houses are built at a higher pace for rapidly growing populations, the standards for new houses are mostly lower and there are no discussions of overall legislation of the type discussed in Europe.

Passive houses could be built using a number of different materials and technologies. The key idea is that all parts of the house, when looked upon as a system, should contribute to the overall goal of energy efficiency.

Box 21.1 Large-Scale Housing Projects – Dongtan and Masdar

China has set the goal of constructing new homes for 400 million people by 2017. Among the many construction projects that are run within this program, the Dongtan project constitutes a commitment to build a city that is completely environmentally efficient. This city is currently being built off the coast of Shanghai. The city will be based on densely built neighborhoods and it will house some 500,000 residents.

The city will be powered by local wind turbine farms and large-scale photovoltaic production, and advanced CHP plants will be fuelled by rice husks to generate the energy that will be necessary to heat buildings.

One other example of a sustainable housing project in the oil-rich Arab emirate of Abu Dhabi is the Masdar project. This city, too, will be a densely populated square-walled city, powered by large photovoltaic plants in the Abu Dhabi desert. In this city cars will be banned, and the city, which will cover 6 square miles, will have a network of shaded pedestrian walkways for transportation on foot in the 120 °F summer heat.[1]

It is possible to build highly energy-efficient houses, using both modern materials and renewable building materials from nature. There are examples of walls that are built with mud and mixed with some concrete, using a technology called "rammed earth"; straw is sometimes used as an insulation material in modern houses. Despite this we must keep in mind that most new houses on the globe are still made of industrial materials and that they are designed to use traditional heating and lighting systems. It will require a massive effort to transform the building sector, to employ new construction materials and methods on a large scale. In this case as well, the transformation needs to be planned and all aspects of the transformation need to be taken care of.

21.3   REFURBISHMENT OF EXISTING HOUSES

Existing houses can be made much more energy-efficient through improved insulation, new heating systems, such as heat pumps, and other measures. This is another aspect of the transformation that will need to be planned in order to reduce energy consumption on a large scale in a short period of time. Subsidies will be needed in order to speed up the process and the need to invest has to be compared to the investment need and savings potential of other possible measures.

Box 21.2 Refurbishment in Colorado

In a competition in Colorado the home of the Culpepper family in Colorado qualified as the most energy-inefficient home of all 10,000 applicants. The family was awarded a complete energy makeover, worth 25,000 dollars.

This retrofit with new energy-efficient installations immediately cut the family gas bills in half, saving 150 dollars per month in the winter. The changes have also made the home more comfortable. It has become significantly warmer in the winter and cooler in the summer.

However, the expert on building energy solutions, Dan Chiras, holds that these amounts do not need to be spent. A few hundred dollars on improved insulation will work wonders for many families and a few thousand will achieve even larger improvements.[2]

21.4   SUSTAINABLE CITY PLANNING

The energy efficiency of transportation within a community is largely dependent on the overall planning and layout of a city. Population density is one aspect that determines energy efficiency, because with increasing density, the whole city can be served by a small public communication network with frequent departures of buses and trains, in a way that has already been described. In the cases of the projects in Dongtan and

Masdar, a high population density is key to the energy efficiency of these future cities and in such cities a large portion of transportation, even in a large city, could be undertaken by foot or by bicycle.

When new districts are planned in a city, or when the plans for a completely new city are laid down, the public communication network can be made into the backbone for communication within a city. This is the case to some extent in European cities, even though many communication networks can be improved through the implementation of high-speed train and bus services, and more lanes for buses and carpooling.

CHAPTER 22

Transforming Agriculture

Agriculture is one of the most important activities in society. It is important because we produce food through agriculture. With a global population of more than 6 billion people, which is growing, and a global food production that has been running surpluses for a long time, which have now started to decline in volume, we need to take the conditions for agriculture into account when we plan the energy transformation.

For many years, industrial agriculture has provided us with a global grain surplus. The use of fertilizers and industrial methods has increased production faster than the growth of the global population has been able to increase consumption. The old fear held by the economist Robert Malthus, that population growth will be more rapid than the production of food, has for a long time been proven wrong. Since 2000, this situation has changed. During the past few years, the annual grain surplus has turned into a deficit and the stores or grain that have been built up are now gradually in decline.[1]

This situation represents a double challenge for the transformation program. While we need to reduce energy consumption, we also need to maintain production volumes. As has already been mentioned, there are experts who estimate that it is possible to transform agriculture so that yields increase by 15–20 percent. This, however, goes against the argument of ecologists that we need to increase the share of ecological farming and in the light of deteriorating growing conditions in some parts of the world increasing average yields may prove to be an unrealistic goal.

22.1   THE ORGANIZATION OF A CHANGE PROGRAM

Agriculture is very different in different parts of the world. Decisions about agricultural practices are made by a huge number of peasants, farmers and managers of agricultural companies and combines across the globe. During the past 50 years great advances have been made in terms of production quantities per hectare, but this development have largely been based on the increasing use of industrial fertilizers and pesticides that are made from oil and natural gas. Ecological farming, in many areas, is free from these chemicals, but requires instead an increasing amount of mechanical treatment of the soil and the use of petroleum is relatively similar for both alternatives.

In a project for reducing the use of water and chemicals in cotton production, which is described below, IKEA, together with the World Wildlife Fund, has developed and spread dramatically improved methods of cotton production. It is a good idea to consider applying these methods on a large scale in the production of a large number of different agricultural products and all over the world. This type of project would comprise three parts, as shown in figure 22.1, namely the development of improved methods, the development of methods and materials for the spreading of these methods on a large scale, and the large-scale training of farmers in these methods.

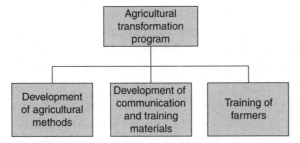

Figure 22.1   Suggested project organization for large-scale managed change project in agriculture

While the above project is organized in order to transform the existing system of agriculture, some authors argue that we need to change the location of agriculture and the organization of farming and food distribution. Following this line of argument, we ought to increase the number of urban farms and farms around cities and to a larger extent eat locally grown crops. If we would like to go through with this type of project, the organization of it would look more like the transformation program concerning the sector of the built environment above. In such a project the subprojects would probably have to deal with regulations and incentives, the development of materials and machinery for large-scale urban farming and the training of homeowners and would-be agricultural entrepreneurs in the practices of this type of farming.

22.2   SMALL-SCALE AND URBAN AGRICULTURE

According to Dale Allen Pfeiffer in *Eating Fossil Fuels*, agriculture accounts for 17 percent of all the energy that is used in the United States.[2] This large amount is used in the forms of diesel, gasoline and electricity, as well as in the forms of fertilizers and pesticides, which are produced using petroleum as raw material.

Agriculture plays a central role in economic development. Through the increased productivity of agriculture we have been able to fuel our past economic growth and free enough production resources, in the form of people who have become available for other jobs. This has been achieved because fewer people are needed to produce food and increasing numbers of people can work in other sectors of the economy. With a reduced share of the population that works in agriculture and with low-value production, we can increase the share that works in service sectors and in high-value production.

It is the access to inexpensive energy that has made it possible for us to maintain our high economic growth. Industrial agriculture has helped us increase the productivity of land by tremendous amounts. But the rapidly growing crops and grain surpluses have stopped growing, and we have depleted the quality of topsoil and the level of aquifers in the process.

Pfeiffer, Heinberg and other authors call for a reform in agriculture, going back to less intensive methods, and a decreased use of fertilizers and pesticides. The measures that they argue in favor of, which will reduce the petroleum use in agriculture, are the following:

- Small-scale, less energy-intensive, farming, based on the model of Cuba after the collapse of the Soviet Union, when this country could no longer receive large imports of oil. The recovery of Cuban agriculture, with a 33 percent increase in the daily food intake to 2470 calories by 2005, compared to the situation in 1994, is described by Pfeiffer as a miracle.
- Ecological agriculture, which creates a sustainable use of land.
- Urban agriculture, which means that an increasing amount of unused urban land could be used for growing vegetables for local consumption. Even many rooftops in cities could be profitably used for this purpose, with reduced energy consumption as a favorable side effect.
- Revitalization of farmer's markets for local agricultural products.

Box 22.1 Added Value – A Brooklyn Farm

The 2.75-acre urban farm and farmer's market, Added Value in Brooklyn, New York, produces vegetables and runs a number of community initiatives:

- Training of youth in how to run a farm and how to grow ecological vegetables.
- Offers education programs in the fields of sustainable business, digital media literacy and community organization.
- Training of teachers in ecology and agriculture.[3]

22.3   IMPROVED EFFICIENCY IN AGRICULTURE

The global furniture retailer IKEA is highly dependent on the access to wood as the raw material for the production of furniture and to cotton and other raw materials for cloth. The company has a set a goal of using wood to 75 percent as the raw material of its furniture. This is because wood is a renewable material, and it is biodegradable. For this reason, IKEA employs ten foresters that inspect forests and methods of forestry in its global supply chain. The foresters work together with the wood suppliers, in order to improve the sustainability of forestry operations.

When in the early years of the new millennium, the question of sustainability of the global cotton production arose within IKEA, the solution that was applied was to copy some of the same methods of cooperation with cotton producers as IKEA applies in its relationship with wood suppliers. The company made a survey of all the cotton-producing areas in the world, and analyzed the different methods of agriculture that were being used. It then, together with the World Wildlife Fund, selected cotton producers in Pakistan for a pilot project, with the aim of implementing better and more sustainable methods of agriculture.

The result of the project was to reduce, by half, the use of water for irrigation, fertilizers and chemical pesticides. The result of this improvement renders, according to IKEA, cotton of better quality than the cotton produced in ecological plantations, and it meets the high standards of quality, applied by IKEA for the cotton that it uses in its furniture and fabrics.

When a Swedish environmental organization invited IKEA and a number of other large Swedish retailers to a discussion on the improvement of the production methods for towels, IKEA told them about its work with the improvement of cotton agriculture, and now this organization has started to learn from IKEA about improvements that can be made.

The improvement of agricultural methods is important, and it can render improvements in other areas than those immediately related to energy consumption. A reduction in the use of fertilizers and pesticides has a direct influence on the use of petroleum in agriculture, since oil is the raw material for most of these products.

In the energy transformation program, it will be important to look for opportunities to improve methods, and it may not be necessary always to make extreme changes to production methods or the scale of production in order to reduce energy consumption. It may be that ecological methods of agriculture are preferable in the long run, but that more rapid results can be achieved through efforts to improve existing methods and technologies.

CHAPTER 23

Behavior and Work Life Change

The purpose of this book is to show the wide range of alternatives that are available in order to create sustainable energy systems for the future. We need to make use of a large number of new technologies, in different areas, that will contribute to the reduction of energy consumption and CO_2 emissions. The development, sales and implementation of these technologies will create a large number of new jobs in development, services (installation, maintenance and spare parts) and management of the new companies that will do business in these new industries. During the life of the transition program in each country, the program itself, and the projects that make up the program, will need managers, administrators and technical experts.

This whole development effort, similar to other previous efforts, will change the private lives and the work lives of many people. It will also contribute a number of opportunities for business development and economic growth.

As a society, we already have much experience of this type of development from a number of large-scale change projects, of which a few, such as the Apollo program and the industrial transformation in the United States during the Second World War, have been used as examples in this book. We also have substantial experience of economic growth, created by increased consumption. We lack experience from large-scale projects aimed at the reduction of consumption and changes that reduce the use of resources without harming the economy.

23.1 THE ORGANIZATION OF A CHANGE PROGRAM

Large-scale change in society requires change on three different levels. The first level is the level of visions and leading ideas that underpin the overall organization and practices in society. One such idea on the highest level is the idea to convert energy systems from fossil to renewable fuels. At a slightly lower level one leading idea is to convert to biogas, and the concrete changes that follow from this idea. The second level is the operations level in companies and in public sector organizations. Changes in ideas on this level are necessary to realize the content of the visions and leading ideas, such as when new vehicles and transportation systems are created based on the ideas of renewable fuels, biogas and the more detailed ideas that follow from this. The third level is the level of citizen and consumer behavior change. When visions and operational development efforts have led to the development of new products, transportation and production systems or simply to the desire to change behavior on a large scale within existing systems, communication efforts directed toward the public are necessary to ascertain the large-scale adoption of new behaviors so that the systems are used, energy savings take place and investments start to pay off.

Many different activities, in terms of the development of ideas, communication and the use of incentives, could be relevant on all three levels above during a change program. The development of visions and leading ideas, operational plans and organizations to realize ideas and the spreading of ideas to the public at large or to segments of the market could take place in a large number of different contexts and it could involve many kinds of activities, ranging from the development of more energy-efficient products and systems, information about energy use in production and distribution and the need to change these systems and some of the routines used in these areas, over propaganda from interest groups to reality television shows that could help people identify change opportunities that could reduce individual energy consumption. This type of development is already under way in society and many times different ideas and initiatives compete with one another. On a very high level, as an example, we could see that the idea put forward in this book that a program of planned and managed change will be necessary is opposed to the view of strong advocates of the free market, who will continue to argue that the market, left on its own, will be able to drive the necessary change at the pace that is needed. The idea of a program for managed change implies that decisions need to be made relatively soon, regarding the contents of possible change programs in a number of areas.

I suggest three high-level subprojects in a program for behavior and work life change, as illustrated by figure 23.1. These subprojects reflect the levels of change above and they are based on the idea that it makes sense for society to decide to support large-scale change of a number of behaviors and that we also need to change visions and business operations accordingly.

Figure 23.1   Suggested project organization for large-scale managed change project in the areas of behavior and work life change

23.2   COMMUNICATION FOR SUSTAINABILITY

In order to achieve large-scale change, decisions for change on the above levels need to be supported by large-scale communication efforts that change visions and leading ideas, operational structures and practices and citizen and consumer behavior. Plans for change need to be based on an understanding of three critical components that are necessary for behavior change, as illustrated by figure 23.2 below.

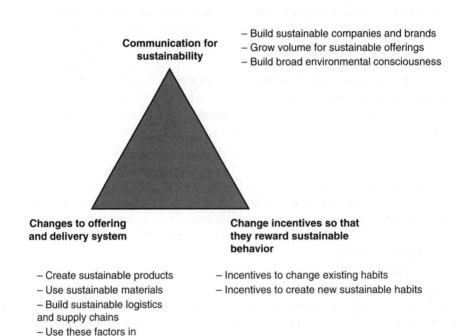

Figure 23.2   Model illustrating relationship between communication of change efforts toward sustainability and the underlying changes to systems and incentives that represent the subjects of communication

The model illustrates the relationship between marketing and communication of sustainable products and ideas, the changes to offerings and delivery systems and incentives that are necessary to make communication credible and to create systems support for behavior change. Discussions of the different aspects of change captured in the model will be found under the relevant headings in other parts of this book.

23.3   THE NEED TO REDUCE CONSUMPTION AND MAINTAIN ECONOMIC GROWTH

It is not an argument in this book that the economy is more important than ecology. The reason for taking the harming of economy into account is that, regardless of what we think, we are all dependent on the economy for the daily supply of food, water, petroleum and electricity. In our relationship to the economy we are like fish that are unaware of the water in which it swims. Most of the time, we are unaware of the economic system. We tend to become aware of it only when it malfunctions, such as when we lose our job during a recession. However, the present economic system is in need of economic growth, in the same way that fish are dependent on oxygen in the water. For the short and medium term, this represents a constraint in terms of the ways and magnitudes that we can afford to change behavior.

Among the advocates of energy efficiency, many argue in favor of reduced consumption, and this alternative is by many taken for granted as the primary alternative for the transition program. The argument is that everybody simply has to understand that we need to reduce our consumption of goods and services, because this consumption drives the need for increased use of energy. If we want to contribute to sustainability, we need to maintain an affluent lifestyle, even though we reduce the number of things we buy.

While advocates of sustainability and energy efficiency argue in favor of a reduction of consumption, the plans of economists and financial planners are based on the idea of continuous economic growth and maintained patterns of consumption. There is a good reason for this. All our economic and financial systems, global as well as national, of all open economies, are based on the expectation of economic growth. The values of pension funds, savings in bank accounts and private and corporate stock ownership are dependent on the expectation by investors that economic growth will continue. This need has gradually been strengthened over the twentieth century, as all mechanisms that peg the value of currencies to each other, or to gold or other valuables, have been abandoned, in the hope and belief that economic growth can continue forever.

In a situation where we need to look forward to a decrease in economic activity among consumers, investors or companies, central banks and government officials would have to re-implement mechanisms into currency systems, which could stabilize the value of money, in the absence of economic growth. This would amount to what economists call a "steady-state economy," which has been advocated by economists such as Professor Herman Daly. The path of transformation from the existing economic system to a steady-state economy on a large scale is somewhat unclear and such a transformation would greatly change our economic landscape. Any attempt to undertake something of this kind would need to be thoroughly planned and debated from a number of perspectives in democratic debates. This would be a long-term process of learning, not only about the possible changes to our economic system, but also about the economic system that we already have.

23.4 THE ROLE OF THE TRANSFORMATION PROGRAM IN BALANCING GROWTH

In this book a situation is described in which we, as a society, will need to invest huge amounts of money in the development of new technologies, and the large-scale implementation of them in energy systems. This investment wave, which will be necessary, will require a large-scale financial commitment from governments, companies, financial institutions and households.

The pressing need to invest in sustainable energy systems will probably require some sacrifices on the part of households and individuals. Many of us will have to prioritize investments in energy improvements in our houses, more energy-efficient cars, and purchase "green" electricity, ecological food and other sustainable products that need to be made available in larger quantities in the coming decades. At the same time, other people may want to cut down on consumption and practices that lead to excess consumption of energy, and which are harmful to the environment. The alternatives of technology development and behavior change could complement each other and different paths for the transition could be chosen by society.

This is only one more example of how we need to do many things at a time in the transformation program. We need to help people reduce consumption in areas that consume large amounts of energy, in order to free financial resources, labor and real estate for activities that are destined to build the new energy-efficient economy. This is another of the management challenges of the next few decades.

The role of the transformation program in balancing this development will be to set goals for change, issue information regarding how the goals could be reached, monitor progress toward these goals and identify and manage deviations from the plan. Even though we have little experience of some aspects of this change, we need to embark upon it and manage it in a professional way, just as we have taken on new challenges in the past.

I do not expect substantial changes in behavior or work life to occur, just because I write about them here. In order to facilitate large-scale structured change, there is a need within the program for a structured stream, information to the public about any changes that they could or should do and a plan to facilitate change and take care of the results of such a change. That means, among other things, plans to constructively make use of the resources that will be freed through this change, in the other streams of this energy transition program.

CHAPTER 24

The Program – A Very Rough Outline

The main argument of the book is that we, in many cases, have too little knowledge about different opportunities for energy improvements. In order to decide about a fully fledged long-range strategy, we need to initiate overall analysis of the transformation challenge and decide on a route forward for such a program. We need to decide about how to handle the restrictions that we have already mentioned:

- Make the best possible use of financial resources.
- Achieve results at the pace that is necessary.
- Make the best possible use of raw materials that are available for energy production.
- Make the best possible use of the land that we have access to.
- Build and utilize competence in the different areas related to the transformation in a planned and structured way.

This means that we need to start the program by making full use of the sustainable technologies that we already have access to, particularly in areas in which activities are financially justifiable for companies. We will also need to make investments for the long term, in systems that are necessary in order to facilitate the transition and for which we can forecast increasing demand in the near future. There are also a number of technologies that need to be developed.

The examples of technologies below are by no means exhaustive. In some cases, the categorizations suggested may even prove wrong. A more detailed analysis will be necessary, in order to make a definitive plan.

24.1 EXISTING TECHNOLOGIES AND PRODUCTS + FINANCIALLY JUSTIFIABLE INVESTMENTS = EXPAND

There are already many technologies in existence that are ready for rapid expansion and market penetration. In many cases, the reason why they do not grow as rapidly as they could is the fact that incumbent technologies, that have been used for a long time, and that are relatively inexpensive, stand in the way of the large-scale application of new alternatives. There exist, however, situations, where investments in improvements of energy efficiency, or the application of new technologies, make perfect sense already.

The opportunities will be discussed area by area.

24.1.1   Transportation

With a price of crude oil above 100 dollars, investments in the production of renewable fuels, that could be mixed with gasoline, and distributed through existing distribution systems, will prove increasingly justifiable from a financial perspective. There is also the opportunity of investing in fuels that require new cars, engines and distribution systems, but for this to happen, there is a need for analysis and overall systems decisions, that will be treated below.

Methanol that is produced from wood and for which there is already a viable large-scale technology in existence is a fuel alternative that could be expanded. Methanol could be mixed into gasoline to a level of around 25–30 percent, without the need of any adjustments in engines. There are also the alternatives of synthetic diesel and biodiesel that could be mixed into diesel and used for trucks, buses and cars. The technologies for production of these alternatives already exist, but only small amounts of these products are available, and the cost is generally higher than for methanol.[1]

In many countries there is also an increasing demand for public regional transportation that could replace some transportation by car, and the investments in expanding capacity on many lines are already financially justifiable. In southern Sweden, regional train companies make substantial investments in new trains, because it is necessary due to increasing demand to increase the number of departures between Malmö and Copenhagen and other cities in the region. The same is true in other parts of Europe and globally and these investments should be sped up in order to increase the capacity of public communications in countries and large city areas where demand is expected to increase.

There are also investments in IT systems for the management of transportation systems for goods and people, that may increase the utilization rate of these systems, and the investments in display and GPS-based information systems at bus and train stations, and at bus stops, that improve the quality of service and make public transportation more attractive. In the same way, marketing campaigns and information campaigns about accessibility and services of public transportation systems and more energy-efficient goods transportation alternatives, such as trains and ships, could increase the use of these systems, encourage switching from more energy-intensive alternatives, and pave the way for the expansion of these systems.

It may also be financially justifiable already, to start new lines in public transportation systems, in order to make the systems denser and more attractive for people who live at a distance from existing lines. This could be done using smaller vehicles for "feeder" lines to the large-scale bus or train lines.

In many cases related to transportation systems, the use of hybrid vehicles or vehicles that could run on alternative fuels or electricity, possibly distributed for large customers through proprietary, or designated, distribution systems, must be considered. It will in many cases be possible to apply new technologies for engines and fuels, at medium scale, for large transportation companies. Biogas is already used by some public transportation companies. In the future, a number of different alternatives, including DME, synthetic diesel, those based on the retrieval of waste, or other alternative fuels, may fuel different transportation systems locally, regionally or nationally, or for particular categories of users and customers.

24.1.2   Energy Production and Distribution

Large-scale wind power programs are already under implementation, and the delivery times for new turbines increase, because of a shortage of capacity. From the point of view of the energy transformation program, we need to rapidly expand the production resources for wind turbines, to the levels that are needed in order to rapidly expand wind power generation. Wind power is, for the moment, the most widely applicable of the technologies for sustainable electricity generation and one of the technologies that currently deliver the lowest production cost. In addition to the expansion of production capacity, a sufficient number of sites for the construction of new wind farms for the coming decades, need to be identified, so that the confidence in this technology is strengthened.

Another alternative that is available for large-scale expansion is cogeneration of heat and electricity. In this case, the long-term access to renewable fuels, such as waste or biogas, needs to be secured.

In the case of geothermal energy and heat generation through heat pumps and solar panels, these technologies are viable at both small and medium scales, for households, offices and factory buildings. Such investments pay back in only a few years and make financial sense.

24.1.3   Industrial Processes

There are many types of energy improvements in industrial processes that can be justified from a financial point of view. Projects that take advantage of these opportunities can be started immediately, but it will, in many cases, take substantial time to realize all opportunities. The reason for this is that it will require substantial amounts of time to manage projects and actually order and install equipment, and sometimes to adapt processes to the new circumstances, and train people in the use and maintenance of new equipment.

In smaller companies, as has already been mentioned, there may not be enough resources to work with the data collection and analysis required in order to decide about alternatives for energy transformation. Therefore, companies need to get together in governmentally financed network activities to get training and consultant support in their transformation efforts. Even if the improvements may be financially justifiable, many small companies will lack the experience of working with systematic change, and many will need support. Many small companies will also lack the financial strength to run transformation efforts, even if the efforts in themselves are financially justifiable over the long term.

24.1.4   The Built Environment

When new housing projects are planned, there are ample opportunities for the construction of highly energy-efficient houses, regardless of the climate conditions of the area where the buildings are going to be erected. There are passive housing concepts and modern building technologies using natural materials, and concepts for the retrieval of energy, and for the application of energy-efficient heating and ventilation systems. The application of these principles makes financial sense already, but the buildings that have been built using these principles will become even more attractive as energy prices increase in the future.

In order to increase the level of knowledge among architects about sustainable building technologies, it will make sense on the level of society to organize large-scale training programs for architects and builders, teaching them the various concepts that are relevant for sustainable housing. The construction industry is otherwise described as slow in its adoption of new principles and technologies "are adopted very slowly." There may be a need both for training and incentives, and, probably, legislation, in order to make sure that the relevant principles are applied as widely as possible.

One other aspect will be the availability of new materials and products that will be in higher demand as construction principles change. It may be necessary to create incentives or subsidized financing alternatives for people and companies which run new housing projects using renewable and energy-efficient materials.

In the case of the refurbishment of existing houses, in order to achieve a higher level of energy efficiency, there are numerous alternatives that are possible. Many of these pay back rapidly, in only a few years, from the application of geothermal or solar heating, to improved insulation, and double, or triple (depending on the climate), glazing. There will be a need for information to households, and owners of existing houses, but this will need to take other forms than in the case of training of architects and builders. In the event that the information and training is successful, there will also be a need to rapidly expand production capacity for the products that will be in high demand.

24.1.5   Behavior and Work Life Changes

There are many simple changes in habits that can have an immediate effect on energy consumption, and on the demand for new capacity from utilities. We could reduce the load at peak levels, simply by switching lights off when we are no longer in a room, and by removing "ghost loads," by switching off television sets and other appliances completely when they are not in use, and by removing battery chargers from sockets when they are not used. The savings in terms of money of such changes in habits will be relatively small for the user, but they may have a substantial total effect on the load on the electricity grid, which is especially important at peak hours, and this may reduce the need to build new capacity if we need to expand our reliance on electricity in the future.

In the same way, companies could replace old computers, copiers and other appliances, for ones that shut down when they are not in use. The same is possible for the use of machinery in production. In many cases, machinery could be installed that will automatically switch off when it is not used.

24.2   EXISTING TECHNOLOGIES AND PRODUCT + NEED TO EXPAND "BOTTLENECK" RESOURCES = EXPAND

In many cases, it will be unclear whether a particular change is financially justifiable or not, but a certain investment will be necessary in order to facilitate some other change. In order, for example, to reduce commuting by car in large city regions, it may be necessary to increase the density of train and bus lines and the frequency of departures in the transportation systems. In some cases, as has been dealt with above, the necessary investment may clearly be financially justifiable, in other cases it may not be justifiable short-term. In order to achieve targets for reduction of the emission of carbon dioxide, and other greenhouse gases, certain investments will be necessary. In the way that was described for the Öresund Bridge between Sweden and Denmark, it will in some cases take some time until commuters have adjusted their patterns of travel and general lifestyles to the new transportation opportunities.

Other savings alternatives may also require a number of different activities in combination. As the number of electric cars and chargeable hybrids increases, the load on the electricity grid will increase. In order to accommodate for this, there will be a need to expand electricity production, possibly launch campaigns to save electricity in households and companies, to reduce load by launching intelligent boilers in homes, freezers and other appliances that could switch off during peak hours. All these and other actions, which may be necessary in order to transfer energy use from petroleum to electricity, may not be financially justifiable in themselves. The overall goal of reducing emissions by switching oil use to renewable energy sources will be so important, that some investments that are not financially justifiable will need to be performed all the same.

24.3 PROMISING TECHNOLOGIES THAT NEED LAST FEW STEPS OF DEVELOPMENT = FINALIZE AND EXPAND IF FINANCIALLY JUSTIFIABLE OR NECESSARY TO REMOVE "BOTTLENECK"

Many small adjustments to existing technologies and products will be possible and necessary in order to make use of new alternative energy sources. One example will be the adaptation of car and truck engines in order to use the new fuel DME, which has, by Volvo, been given a very favorable evaluation, as seen from seven different perspectives. The production of DME is not very complicated, and it may be possible to start production within a relatively short period of time, but it will take time to increase capacity and to launch trucks that will be able to use this fuel.

There may be other technologies that could be developed within the next few years, if only sufficient resources are assigned to the project, and a tight plan is developed. In addition to technologies that are already available, we will, for the medium term, need both additional technologies for electricity production and new engine and fuel technologies for transportation. The demand for electricity, as has been indicated above, will vary depending on how we choose to solve our future transportation needs.

Examples of possible new technologies for electricity generation that could be developed to economic viability during the next few years could be wave energy. It may be the case that the cost of photovoltaic panels could decrease substantially with increasing demand. All the different opportunities to reduce energy consumption and to transfer energy use to renewable sources need to be evaluated and analyzed both from the point of view of investment needs, cost per unit of energy, generation capacity and a number of other factors.

24.4   TECHNOLOGIES WITH MAINLY LONG-TERM POTENTIAL = EVALUATE NEED AND DEVELOP STRATEGY

Some technologies will not deliver one single kilowatt-hour of energy within the next decade or two. This could either be because they are still at an early stage of technology development, because there are a number of difficult technology issues that remain to be solved, or because they require large-scale projects in order for energy production to be possible. The latter is true for the construction of large-scale tide energy dams.

When the above alternatives have been analyzed, and the long-term need for additional energy sources has been assessed, such alternatives may be included in the plan. At this point we cannot exclude any particular energy technologies.

24.5   VISIONARY TECHNOLOGIES AND POSSIBLE "SHORTCUTS" = TRY SOME AT SMALL SCALE + EVALUATE + CONSIDER RAPID EXPANSION IF SUCCESSFUL

Visionary technologies that may offer substantial new, and unexpected, opportunities have already been briefly discussed. It needs also to be emphasized that many possible solutions, not only to energy technologies, but also to various production issues and other technical problems, may have been overlooked. The opportunities to make new and radical innovations have not been exhausted.

It will be justifiable to include the development of a number of visionary new technologies into the plan for the energy transition program. This development may be subsidized through large-scale development projects, in the same way that other opportunities could be subsidized.

Conclusion

The market-based approach to energy systems transformation, which is currently being pursued, will not suffice to create the large-scale transformation that we need. All the necessary financial resources will not be provided by the market on a timely and voluntary basis. A number of planned measures will be necessary in order to facilitate the necessary investments for the change program.

Energy systems are, and the transformation of them, would be, in general, more complex than many people realize. The complexity of these systems and the different ways by which changes in one system influence the need for change in other systems require analysis. In order to make sure that the money invested in development and transformation activities gives the most rapid results possible we need to make a strategy and a plan for the overall transition, and break these down into strategies and plans for the different streams of the program, and the projects that would need to be included in each stream. Each stream and project, and the program itself, in a particular country will need a steering committee, a project management structure and project resources. In a planned program, which is run based on a plan that should be supported by the government of a country, a set of tools for program and project management are needed.

This type of program will require substantial investments from companies, and from households and individuals. Some of this investment will be provided through normal market-based transactions, and some will have to be provided through government funds, in the forms of direct investment, subsidies, price controls and other financial mechanisms. At this point, we do not know how much resources can be provided by the market, and how much that will need to be funded by the government. This has to be analyzed as part of the strategy analysis within the program.

Government funding and planning of investments and change activities does not imply socialism, or a move by the United States, or any other country, away from the market economy system. There are already ample examples throughout the twentieth-century US history, and before, of the use of large-scale planned industrial transformation in order to take on important challenges. The examples of the transformation of the entire American industry in order for the United States to become "the arsenal of democracy" and lead the Allied countries in an effort to win the Second World War, The Marshall Plan, and the Apollo program support the argument that planned efforts have been an important aspect of global economic life for a long time. Furthermore, these efforts have not only contributed substantially to economic growth. They have also, in some cases, made possible the development of technologies that would not have seen the light of day without government funding. This is due to the long development cycles of general purpose technologies that sometimes span several decades, and the uncertainty of such investments, which cannot be handled by the market mechanism left on its own.

All things taken together, many things speak to the advantage of planned change and partial government funding. The most compelling economic reason for change will not be the cost incurred by extreme weather conditions that have been experienced during the last couple of years, or the climate change, sometimes referred to as "global warming" that we are also starting to experience. It will be the increasing prices of oil and other energy resources that will be one of the consequences of the peak in oil production that many experts now hold to be imminent, or that may already have passed; this will turn out to be the factor that motivates the financial markets of the world to provide some of the resources that will be necessary for the energy transition. In order for companies to do their part of the transformation, companies need to know the long-term policy for the transition. In the words of the CEO of Volvo, Leif Johansson: "We are obviously a part of the problem.... In order for the effort to develop carbon dioxide free transportation to become successful, and in order for us to succeed in our endeavour to become part of the solution, broad high level agreements will be necessary" (Volvo AB booklet – Med klimatfrågan i fokus, Gothenburg 2007).

In order for change to take place, a large number of people need to take on the challenge of learning about the transformation needs and the different alternatives that are at our disposal. We need to start to build the competencies that will be necessary, and organize the competencies in a well-structured and managed program, streams and project teams. We need to start immediately with the structured efforts.

Notes

1 Executive Summary

1. Heinberg, Richard – *The Party's Over*, p. 157.
2. Ibid., pp. 132–137.
3. Blinge, M. – "Transportsektorns alkoholberoende måste brytas", article in Johansson, B. *Bioenergy –To What and How Much?*.
4. Boyle, Godfrey – *Renewable Energy*, p. 285.

2 Who Should Read This Book and Why?

1. Obama, B. (2006) *The Audacity of Hope*, Vintage Books, New York, pp. 49–50

3 The Sustainability Challenge

1. Larsson, Mats – *The Limits of Business Development and Economic Growth*.
2. Tertzakian, Peter – *A Thousand Barrels a Second*, McGraw-Hill, New York 2005, p. 62.
3. Ruttan, Vernon – *Is War Necessary for Economic Growth?*

4 Four Scenarios for the Future

1. Heinberg, Richard – *The Party's Over*, p. 157.
2. Boyd, Godfrey – *Renewable Energy*, p. 285.
3. Heinberg, Richard – *The Party's Over*, pp. 132–137.
4. Heinberg, Richard – *Power Down*, p. 21.
5. Darley, Julian – *High Noon for Natural Gas*, Chelsea Green Publishing Company, River Junction 2004, p. 2.
6. Deffeyes, K – *Beyond Oil*, Hill & Wang 2005, p. 3.
7. The Royal Swedish Academy of Sciences – Statements on Oil, 2005.
8. Simmons, Mathew – *Twilight in the Desert*, pp. 151–180.
9. The Royal Swedish Academy of Sciences – Statements on Oil, 2005.
10. Darley, Julian – *High Noon for Natural Gas*, Chelsea Green Publishing Company, River Junction 2004, p. 5.
11. Hirsch, Robert (ed.) – *Peaking of World Oil Production: Impact, Mitigation and Risk Management*.

5 Three Examples of Large-Scale Transformation Efforts from American Twentieth-Century History

1. Gordon, John Steele – *An Empire of Wealth*, p. 350.
2. Ibid., pp. 353–354.
3. Behrman, G. – *The Most Noble Adventure*, pp. 22–23.
4. Ibid.
5. Ibid., p. 4.
6. Ibid., pp. 177–179.
7. Ibid., pp. 108–109.
8. Ibid., pp. 181–182.

9. Ibid., pp. 138–139.
10. Ibid., p. 333.
11. Orloff, R.W. & Harland, D.M. – *Apollo: The Definitive Source Book*, Springer, Berlin 2006, p. 27.
12. Collins, J.D. & Porras, J.I. – *Built to Last*, Harper, New York, pp. 82–85.

6   Gradual Change Has Caused the Need for Energy Transformation

1. Heinberg, R. – *The Party's Over*, p. 27.
2. Stalk, George – *Time – The Next Source of Competitive Advantage*, Harvard Business Review, July–August 1988.
3. Womack et al. – *The Machine That Changed the World*, p. 83.
4. Krugman, Paul – *Peddling Prosperity*, pp. 3–15.

7   Thoughts on Planning and Market Economics

1. Senge, Peter – *The Fifth Discipline*, pp. 22–23.
2. Tainter, Joseph – *The Collapse of Complex Societies*.
3. Tertzakian, Peter – *A Thousand Barrels a Minute*, pp. 12–22.
4. Hirsch, Robert (ed.) – *Peaking of World Oil Production: Impact, Mitigation and Risk Management*.
5. Krugman, P. – *Peddling Prosperity*, pp. 109–112.
6. Acemoglu, D., Johnson, S. & Robinson, J. (2004). *Institutions as the Fundamental Cause of Long Run Growth*. NBER Working Paper Series: 10481, NBER, Cambridge Mass.
7. De Long, J.B. (1991). *Productivity Growth and Investment in Equipment: A Very Long Run Look*. Harvard University and NBER, Cambridge Mass.
8. The concept of "creative destruction" in economics was developed by the Austrian economist Joseph Schumpeter in the 1930s.

9   Tools for Program and Project Executive Direction

1. Senge, P. – *The Fifth Discipline*, pp. 163–252.
2. Collins, J.C. & Porras, J.I. – *Built to Last*, Harper Collins, New York 1994.

10   The Standard Tools of Program and Project Management

1. Blinge, M. – "Transportsektorns alkoholberoende måste brytas", article in Johansson, B. *Bioenergy – To What and How Much?*

12   New Tools for Learning and Analysis

1. Surowiecki, J. – *The Wisdom of Crowds*, Little, Brown, New York 2004, pp. xx–xxi.
2. Westling, G. – *Balancing Innovation and Control: The Role of Face-to-Face Meetings in Complex Product Development Projects*, EFI, The Economic Research Institute, Stockholm School of Economics, 2002.
3. Wolfers, J. & Sitzewitz, E. – *Prediction Markets*, Working Paper 10504, National Bureau of Economic Research, Cambridge MA 2004.
4. Ibid.
5. Chen, K.-Y. & Plott, C.R. – *Information Aggregation Mechanisms – Concept, Design and Implementation for a Sales Forecasting Problem*, Social Science Working Paper 1131, California Institute of Technology (Caltech) 2002.

13 Financial Tools

1. Komor, P. – *Renewable Energy Policy*, iUniverse, New York 2004, pp. 109–115.
2. Ibid., pp. 84–87.
3. Lietaer, B. – *The Future of Money*, p. 201.
4. Ibid.
5. Ibid., pp. 198–200.

14 Change Happens in Steps

1. The e-step model is now available in English in "The Limits of Business Development and Economic Growth" by Mats Larsson (Palgrave Macmillan 2004).

15 A Step Change Model for Energy

1. Ruttan, Vernon – *Is War Necessary for Economic Growth?*

16 Change in Different Sectors of the Economy

1. Collins, J. & Porras, J. J – *Built to Last*, p. 46.

17 The Transformation Program

1. Heinberg, R. – *Powerdown*, p. 19.
2. The Royal Swedish Academy of Sciences – Statements on Oil, Stockholm 2005.
3. Godfrey Boyle – *Renewable Energy*, The Open University 2004.

18 Transforming Transportation

1. Heinberg, Richard – *The Oil Depletion Protocol*, p. 41.
2. Bergkvist, Ann-Kristin & Lindmark, Magnus – "Satsa på radikal teknikförändring, inte på biobränslen", essay contained in the book Johansson, B. (Ed.) – *Bioenergy – For What and How Much?*, Formas, Stockholm 2007.
3. Rydberg, Torbjörn – "Dagens analysmetoder skapar övertro på bioenergi", essay contained in the book Johansson, B. (Ed.) *Bioenergy – For What and How Much?*, Formas, Stockholm 2007.
4. Berndes, Göran; Hansson, Julia & Wirsenius, Stefan – "Biomassa – en knapp resurs i ett globalt perspektiv", essay in the book Johansson, B. (Ed.) *Bioenergy – For What and How Much?*, Formas, Stockholm 2007.
5. Blinge, Magnus – "Transportsektorns alkoholberoende måste brytas", essay in the book Johansson, B. *Bioenergy – For What and How Much?*, Formas, Stockholm 2007.
6. Volvo – "Med klimatfrågan i fokus", booklet published in the spring of 2007.
7. What is Enlightenment? Issue 38, October–December 2007.
8. Ibid.
9. Kidd, J.B. & Stumm, M. – *High Speed Maglev Logistics. Is This the "Golden Goose" We Are Looking For?*
10. Park and Ride means that people are offered the opportunity to park their cars close to a railway station or a bus stop, and use the public transportation network for the distance into a city center, or to an amusement park, exhibition center, or other places of public interest.

19   Transforming Heating and Utilities

1. Boyle, G. – *Renewable Energy*, p. 285.
2. Ibid., p. 208.
3. Ibid., p. 196.
4. Ibid., p. 298.
5. Ibid., p. 342.
6. Ibid., p. 348.

20   Transforming Industrial Processes

1. WWF & ETNO – Saving the Climate @ the Speed of Light.

21   Transforming the Built Environment

1. What Is Enlightenment? Issue 38, October–December 2007.
2. Chiras, D. – *The Homeowner's Guide to Renewable Energy*, p. 1.

22   Transforming Agriculture

1. Heinberg, R. – *Powerdown*.
2. Pfeiffer, D.A. – *Eating Fossil Fuels*, p. 19.
3. http://www.added-value.org/initiatives.php

24   The Program – A Very Rough Outline

1. Volvo AB booklet – "Med klimatfrågan i fokus", 2007.

References

Acemoglu, D., Johnson, S. & Robinson, J. – *Institutions as the Fundamental Cause of Long Run Growth*, NBER Working Paper, Series: 10481, NBER, Cambridge Mass 2004.

Behrman, G. – *The Most Noble Adventure*, Free Press, New York 2007.

Boyle, G. (ed.) – *Renewable Energy*, Oxford University Press, Oxford 2004.

Boyle, G., Everett, B. & Ramage, J. – *Energy Systems and Sustainability*, The Open University, Oxford 2003.

Bromberg, J.L. – *NASA and the Space Industry*, Johns Hopkins, Baltimore 1999.

Brown, J. & Isaacs, D. – *The World Café*, Berret Koehler, San Francisco 2005.

Brown, S.L. & Eisenhart, K.M. – *Competing on the Edge*, Harvard Business School Press, Cambridge 1998.

Chiras, D. – *The Homeowner's Guide to Renewable Energy*, New Society Publishers, Gabriola Island 2006.

Christopher, M. – *Logistics and Supply Chain Management*, Pearson Education Ltd., New York 2005.

Capehart, B.L., Turner, W.C. & Kennedy, W.J. – *Guide to Energy Management*, The Fairmont Press, Lilburn 2006.

Daly, H.E. – *Beyond Growth*, Beacon Press, Boston 1996.

Deffeyes, K.S. – *Beyond Oil*, Hill and Wang, New York 2005.

De Long, J.B. – *Productivity Growth and Investment in Equipment: A Very Long Run Look*, Harvard University and NBER, Cambridge, Mass. 1991.

Etsy, D. & Winston, A.S. – *Green to Gold*, John Wiley & Sons, New York 2006.

Gordon, J.S. – *An Empire of Wealth*, Harper Perennial, New York 2004.

Grant, L. – *The Collapsing Bubble*, Seven Locks Press, Santa Ana 2005.

Greco, T.H. – *Money*, Chelsea Green Publishing, White River Junction 2001.

Hacker, B.C. – *American Military Technology*, Johns Hopkins, Baltimore 2006.

Heinberg, R. – *The Party's Over*, New Society Publishers, Gabriola Island 2003.

Heinberg, R. – *Power Down*, New Society Publishers, Gabriola Island 2004.

Heinberg, R. – *The Oil Depletion Protocol*, New Society Publishers, Gabriola Island 2006.

Hirsch, Robert (ed.) – *Peaking of World Oil Production: Impact, Mitigation and Risk Management*.Projectcensored.org/newsflash/the_hirsch_report.pdf. 2005.

Howe, J.G. – *The End of Fossil Energy*, McIntire, Waterford 2003.

Inslee, J. & Hendricks, B. – *Apollo's Fire*, Island Press, Washington 2008.

Johansson, B. (ed.) – *Bioenergy*, Formas, Stockholm 2007.

Johnson, S.B. – *The Secret of Apollo*, Johns Hopkins, Baltimore 2002.

Keynes, J.M. – *The General Theory of Employment, Interest and Money*, Prometheus Books, New York 1997.

Komor, P. – *Renewable Energy Policy*, iUniverse, New York 1004.

Krugman, P. – *Peddling Proseprity*, Norton & Co., New York 1994.

Kunstler, J.H. – *The Long Emergency*,Atlantic Monthly Press, New York 2005.

Leeb, S. – *The Coming Economic Collapse*, Warner Business Books, New York 2006.

Lietaer, B. – *The Future of Money*, Random House, New York 2001.

Lietaer, B. & Belgin, S.M. – *Of Human Wealth*, Galley Edition, version 2.11, Human Wealth Books and Talks, Boulder Colorado 2004.

Liker, J.K. – *The Toyota Way*, McGraw-Hill, New York 2004.

McBay, A. – *Peak Oil Survival*, The Lyons Press, Guilford 2006.

McCurdy, H.E. – *Faster Better Cheaper*, Johns Hopkins, Baltimore 2001.

Murray, C. & Bly Cox, C. – *Apollo*, South Mountain Books, Burkittsville 2004.

Orloff, R.W. & Harland, D.M. – *Apollo*, Praxis Books, Chichester 2006.

Peters, T.J. & Waterman, R.H. – *In Search of Excellence*, Harper & Row, New York 1982.

Pfeiffer, D.A. – *Eating Fossil Fuels*, New Society Publishers, Gabriola Island 2006.

Ruttan, V.W. – *Is War Necessary for Economic Growth?* Oxford University Press, Oxford 2006.

Senge, P. – *The Fifth Discipline*, Currency Doubleday, New York 1990.

Shenhar, A.J. & Dvir, D. – *Reinventing Project Management*, Harvard University Press, Cambridge 2007.

Simmons, M.R. – *Twilight in the Desert*, JohnWiley & Sons, New York 2005.

Smith, A. – *The Wealth of Nations*, Penguin Books, London 1997.

Surowietzky, J. – *The Wisdom of Crowds*, Anchor Press, New York 2005.

Tainter, J. – *The Collapse of Complex Societies*, Cambridge University Press, Cambridge 1988.

Tertzakian, P. – *A Thousand Barrels a Second*, McGraw-Hill, New York 2006.

Womack, J.P., Jones, D.T. & Roos, D. – *The Machinve That Changed the World*, Harper Collins 1990.

Worldwatch Institute – *Biofuels for Transport*, Earthscan, London 2007.

Index

Note: Page numbers with notes are given with "n." followed by the note number. If there are two notes in the same page the chapter number has also been given in brackets for reference.